T0205834

Ion Exchange and Solvent Extraction

A Series of Advances

Volume 22

ION EXCHANGE AND SOLVENT EXTRACTION SERIES

Series editors
Arup K. Sengupta
Bruce A. Moyer

Founding Editors
Jacob A Marinsky
Yizhak Marcus

Contents of Other Volumes

Ion Exchange and Solvent Extraction

A Series of Advances

Volume 22

Edited by

ARUP K. SENGUPTA

Lehigh University, Bethlehem,
Pennsylvania, USA

CRC Press

Taylor & Francis Group
Boca Raton London New York

CRC Press is an imprint of the
Taylor & Francis Group, an **informa** business

CRC Press
Taylor & Francis Group
6000 Broken Sound Parkway NW, Suite 300
Boca Raton, FL 33487-2742

First issued in paperback 2022

© 2016 by Taylor & Francis Group, LLC
CRC Press is an imprint of Taylor & Francis Group, an Informa business

No claim to original U.S. Government works

ISBN 13: 978-1-4987-6122-2 (hbk)
ISBN 13: 978-1-03-240252-9 (pbk)

DOI: 10.1201/b20021

Visit the Taylor & Francis Web site at
http://www.taylorandfrancis.com

and the CRC Press Web site at
http://www.crcpress.com

Contents

Preface

No ingenuity grows in isolation. The true versatility and usefulness of a scientific field lies in its ability to connect to others for enhanced synergy. During the six decades following the Second World War, the field of ion exchange has permeated into a wide array of applications ranging from mining to microelectronics, environment to energy, drug delivery to detection, food to fertilizer, chemical cleaning to catalysis, bioseparation to brackish water, and many others. As environment and energy-related regulations grew, ion exchange took a dominant role in offering solutions to many concurrent problems both in the developed and the developing countries. New materials and new processes are rapidly growing. Ion exchange fundamentals and its application opportunities provide a platform welcoming a diverse group of professionals from chemistry, chemical engineering, environmental engineering, polymer chemistry, physical chemistry, biotechnology, and others. The six chapters I have assembled in this latest volume reflect diverse contributions from researchers across the globe who are making noticeable strides in the field of ion exchange.

In the coming years, desalination will increasingly become a more attractive option for the production of drinking water. For inland brackish water desalination plants like those in Texas and Colorado, the disposal of waste brine constitutes a significant portion of the plant-operating costs. Increasing the recovery of the plant would decrease the volume of waste brine, but this is not immediately possible due to the potential for the formation of $CaSO_4$ scale. Chapter 1 describes how a mixed bed of ion exchange resins may be used to remove sulfate from the feed water of desalination plants, thereby preventing the formation of $CaSO_4$ and increasing the recovery of the desalination process. The self-regenerating hybrid ion exchange–reverse osmosis process relies on the ability of ion exchange resins to allow the selective removal of sulfate and replacement by nonprecipitate forming chloride. The water is then desalinated at increased recovery without threat of sulfate scaling. The waste chloride brine from the desalination process is then used to regenerate the ion exchange; no additional chemical inputs are necessary. A tunable bed of mixed anion exchange resins forms the heart of the process. The hybrid methods may be adapted to suit any feedwater composition merely by changing the type of resins used or their mixing ratios.

While some water sources may be too saline for human consumption, other water sources may contain trace contaminants that are unsuitable for use in agriculture. Boron, for example, is a ubiquitous, naturally occurring element that may exist as a variety of different complex compounds. However, water containing boron at concentrations above 1 mg/L will inhibit plant growth and cause boron poisoning. Unfortunately, most common methods of water treatment are unable to effectively reduce the concentration of boron, and even advanced methods, like reverse osmosis, are ineffective at reducing boron concentrations without significant alkali dosing. A solution to this issue is presented in Chapter 2, where methods of boron removal using ion exchange are detailed. Ion exchange sorbents with unique functional groups may be used to selectively remove boron using either standard sorption methods or hybrid processes and may provide more efficient boron removal than other commonly used methods.

Besides contaminant removal, ion exchange may also serve as a catalytic material. Within the chemical synthesis industry, alkene epoxidation is useful for creating highly reactive compounds that have applications in a wide variety of industries. However, the underlying chemistry of epoxidation uses environmentally unfriendly oxidizing agents, such as peracetic acid, and also produces hazardous waste streams containing acid waste or chlorinated by-products. Previous research has demonstrated that molybdenum (IV) may, instead, be used as a catalyst for epoxidation but was not economically viable for industrial applications due to increased purification requirements. Previous research using batch studies has shown that polymer-supported molybdenum complexes may serve as effective catalysts while avoiding excess post-production purification steps, but, to date, there have been no efforts in transforming the process to a continuous flow process. Chapter 3 details an extensive study on the preparation and characterization of the molybdenum catalyst, an assessment of the catalytic activity and reusability of the catalysts, and the efficiency of using a FlowSyn reactor, a continuous flow epoxidation system.

Removing toxic heavy metals from the background of waste streams containing suspended solids or from the soil of hazardous waste sites is a difficult task. Metals can exist in a variety of states: as a solid precipitate with an inert solid phase, as a trace contaminant in the background of a buffer, or co-existing with other toxic metals bound to the soil through ion exchange, and each case presents its own set of challenges for removal. Ion exchange resins with specific chelating functional groups are a robust and reliable method for the selective removal of metals from water. However, their physical characteristics make them inadequate for similar applications for the removal from solids. To this end, Chapter 4 provides a detailed investigation on two new ion exchange materials, a thin-sheet composite ion-exchange material (CIM) and an ion exchange fiber (IXF), which can be used to selectively remove contaminants from heavy metal–laded sludge/slurry. The CIM is a microporous composite sheet of ion exchange powder embedded in polytetrafluoroethylene, which is amenable for a continuous sorption/regeneration process. Meanwhile, the IXF is a thin strand of ion exchange resin that has an extremely high affinity toward metals while also possessing fast kinetics for removal.

Ion exchange resins may also serve as a means for separating different dissolved species from concentrated waste brine streams using the method of acid retardation. The process relies on the fact that acids travel slower through a column of anion exchange resin than metallic species. However, the efficiency of the acid retardation can be reduced due to mixing from longitudinal dispersion effects. These effects are normally diminished by passing the brine solution upward while passing regenerant solution downward, but due to the shrinking and swelling of the resin, dispersion effects cannot be avoided on the large scale. Chapter 5 introduces a new method for reducing the effects of longitudinal dispersion by filling the pore space between resin beads with an organic liquid that is immiscible in water. This creates a thin film of water around the ion exchange beads, reducing the total volume of water inside the column but still allowing ion exchange to occur. This method has no effect on ion exchange selectivity but greatly increases the kinetics of removal and facilitates chromatographic elution of different species that would normally be inseparable.

For developing countries, checking for the presence of metal contaminants may not be immediately possible without advanced testing equipment. Other commonly practiced methods require some type of sample preparation or experience interference from background species. Chapter 6 presents research regarding a new hybrid inorganic material (HIM) capable of detecting the presence of trace metal contaminants using pH as a surrogate indicator. When passing solution through a column containing the calcium–magnesium–silicate material, the effluent will be at a consistent pH in the presence of common background ions. But, in the presence of toxic heavy metals like copper or nickel, the effluent pH from the HIM column changes to several pH units lower, giving a quick and reliable means for determining the presence of contaminants. The user only has to measure effluent pH using a pH probe or observe color using an indicator solution or color.

As editor of the series, I am immensely thankful to all the chapter contributors of this 22nd volume. Last but not least, I acknowledge with thanks a great deal of assistance and unselfish effort of Ryan Smith and Mike German, doctoral students at Lehigh University, for attending to the many details that brought this volume to its successful closure.

Arup K. SenGupta
Lehigh University
Bethlehem, Pennsylvania

Editor

Arup K. SenGupta is currently a P.C. Rossin Professor of the Department of Civil and Environmental Engineering and the Department of Chemical Engineering at Lehigh University, Bethlehem, Pennsylvania. He received his bachelor's degree in chemical engineering from Jadavpur University, Kolkata, India, in 1973, worked with Development Consultants Pvt. Ltd., in India, as a process development engineer from 1973 to 1980, and in 1984, he obtained a PhD in environmental engineering from the University of Houston, Houston, Texas. In 1985, he joined Lehigh University. He served as the editor of the *Reactive and Functional Polymers* journal from 1996 to 2006.

Dr. SenGupta's research encompasses multiple areas of separation and environmental processes including ion exchange. For his research, Dr. SenGupta has received a variety of awards including the 2009 Lawrence K. Cecil Award from the American Institute of Chemical Engineers (AIChE), the 2009 Astellas Foundation Award from the American Chemical Society (ACS), the 2007 Grainger Silver Prize Award from the National Academy of Engineering (NAE), the 2004 International Ion Exchange Award at the Cambridge University in England, to name a few. He has 10 US patents (8 awarded and 2 pending) mostly in the area of ion exchange science and technology.

For his research inventions and their impact on both the developed and the developing countries, Dr. SenGupta has been inducted as a fellow of the National Academy of Inventors (NAI), AIChE, and the American Society of Civil Engineers (ASCE).

Contributors

Prasun K. Chatterjee
CTO, WIST Inc.,
Brighton, Massachusetts

Alex N. Gruzdeva
Vernadsky Institute
Russian Academy of Sciences
Moscow, Russia

Enver Guler
Nanotechnology Research and
Application Center (SUNUM)
Sabanci University
Istanbul, Turkey

Idil Yilmaz Ipek
Chemical Engineering Department
Faculty of Engineering
Ege University
Izmir, Turkey

Nalan Kabay
Chemical Engineering Department
Faculty of Engineering
Ege University
Izmir, Turkey

Ruslan Kh. Khamizov
Vernadsky Institute
Russian Academy of Sciences
Moscow, Russia

Sultan Kh. Khamizov
NewChem Technology, LLC
Moscow, Russia

Anna N. Krachak
Vernadsky Institute
Russian Academy of Sciences
Moscow, Russia

Misbahu Ladan Mohammed
Centre for Green Process Engineering
School of Engineering
London South Bank University
London, UK

Tabish Nawaz
Civil and Environmental Engineering
Department
University of Massachusetts
Dartmouth, Massachusetts

Basudeb Saha
Centre for Green Process Engineering
School of Engineering
London South Bank University
London, UK

Arup K. SenGupta
Department of Civil and Environmental
Engineering
Lehigh University
Bethlehem, Pennsylvania

Sukalyan Sengupta
Civil and Environmental Engineering
Department
University of Massachusetts
Dartmouth, Massachusetts

Ryan C. Smith
Department of Civil and Environmental
Engineering
Lehigh University
Bethlehem, Pennsylvania

Natalya S. Vlasovskikh
NewChem Technology, LLC
Moscow, Russia

Mithat Yuksel
Chemical Engineering Department
Faculty of Engineering
Ege University
Izmir, Turkey

1 Integrating Tunable Anion Exchange with Reverse Osmosis for Enhanced Recovery during Inland Brackish Water Desalination

Ryan C. Smith and Arup K. SenGupta

CONTENTS

For inland desalination plants, managing and discarding produced brine leftover can involve significant operating costs. By increasing the recovery of the desalination process, brine volume and disposal costs can be reduced. However, achieving higher recovery is not immediately viable, as the process involved has a higher potential for calcium sulfate precipitation, which, during reverse osmosis (RO) processes, can foul and eventually damage the RO membrane.

Ion exchange may be used as a pretreatment method to selectively remove and replace sulfate by chloride to eradicate any threat of fouling. The RO process can then be operated at higher recoveries without any threat of sulfate scaling because of the removal of sulfate by the ion exchange column. After RO, the leftover concentrate, highly concentrated chloride brine, can be used as a regenerant for the ion exchange column without requiring the purchase of additional chemical regenerant. By changing the type and/or mixing together characteristically different ion exchange resins, the selectivity of the ion exchange column can be precisely tuned to remove sulfate regardless of the feedwater composition.

Results demonstrate that a properly designed hybrid ion exchange–reverse osmosis system can effectively eliminate the potential for CaSO$_4$ scaling sustainably without requiring external regenerant. The selectivity of the ion exchange resin has a significant role in controlling sulfate removal, and it is possible to precisely predict how the resin selectivity changes depending on the solution composition or mixing ratio with another resin.

1.1 INTRODUCTION

1.1.1 Brackish Water Desalination in the United States

Throughout the United States, 71% of the population receives its drinking water from surface water sources such as lakes and streams.[1] However, in recent years, surface and groundwater resources have been declining.[2,3] Because of anthropogenic climate change, temperatures in arid regions such as the US Southwest have been increasing, resulting in reductions in precipitation. In the future, water availability in this region will decline.[4] As a result, the desalination of previously untapped saline water sources is now being considered as an option for supplying water to arid regions.[5-8] In these cases, standard methods of drinking water treatment are unable to reduce the total dissolved solids (TDS) content enough for human consumption and advanced desalination treatment methods are required.[9] As of 2006, the United States produces approximately 5.6 million m^3/day of drinking water by desalination.[10] Currently, there are approximately 250 desalination plants operating within the United States with most located in Florida, California, and Texas.[11] Of all the desalination facilities in the United States, 65% use brackish water sources. Brackish water refers to water that has a TDS content of 500–10,000 mg/L.[12]

The most common method of desalination in the United States is reverse osmosis (RO).[9] A semipermeable RO membrane is used to physically separate pure water from dissolved ions in solution. The solution that passes through the membrane is referred to as permeate while any remaining saline solution is referred to as concentrate. For brackish water desalination plants, the management of excess volumes of leftover concentrate constitutes a significant problem as there is no easy method of concentrate disposal.[13] The costs associated with concentrate disposal can contribute to a significant portion, in some cases up to 50%, of the operating costs of the desalination plant.[9] Increasing the recovery of the RO process, even by a small amount, could result in a large reduction in the volume of concentrate produced. For example, increasing the recovery of an RO plant from 80% to 90% would result in a 50% decrease in the volume of concentrate produced. This reduction would not only help reduce the operating costs but would also decrease the environmental impact because of the lower volume of discharged brine.

1.1.1.1 Concentrate Management Strategies

There are several commonly practiced methods of concentrate disposal: discharge to surface water, sewer disposal, deep well injection, evaporation ponds, and land application.[14] Each method has its own drawbacks, and the disposal choice for a municipality is largely dictated by geographical location and plant size.[15] Considering a large brackish water desalination plant, greater than 6 million gallons per day, sewer disposal, evaporation ponds, and land application are not viable options even under normal operating conditions.[14] Sewer disposal is possible only if the receiving wastewater treatment plant grants permission, evaporation ponds are prohibitively expensive because of the large land requirements, and the high salinity and volume of produced concentrate makes land disposal impossible. In arid regions, such as the Southwest United States, sites for surface water discharge are not immediately available, leaving deep well injection as the only possible solution.

Concentrate produced by RO desalination is classified as industrial waste as part of the industrial classification codes used by the United States Environmental Protection Agency (USEPA) and therefore requires a Class I well for disposal.[14]

TABLE 1.1

Commonly Dosed Chemicals/Antiscalants

Additive	Average Dosing[28] (mg/L)
Sulfuric acid	50
Sodium hexametaphosphate	6
Polyacrylic acid	3
Phosphonate	2

Regulations require stricter monitoring and other preventative measures to ensure that the discharged waste does not contaminate any local drinking water aquifers, but do not stipulate what quality of water can be injected into the well. Therefore, if measures are taken to reduce the volume of water produced resulting in an increased salinity, currently practiced disposal methods will not need to be changed.

1.1.1.2 Scaling Prevention Measures

During normal RO operation, the concentration of solution at the surface of the membrane is several times more concentrated than the bulk solution because of the phenomenon of concentration polarization. Precipitates such as $CaCO_3$, silica, or $CaSO_4$ tend to precipitate on the surface of the membrane resulting in membrane scaling and fouling.[16–26] In order to prevent and inhibit their formation, antiscaling chemicals are dosed in the feedwater. Prevention of carbonate and silicate scaling can be controlled through feedwater pretreatment by acid dosing or chemical precipitation.[27] For high sulfate feedwaters, chemical precipitation is not feasible for preventing $CaSO_4$ and acid dosing has little effect.[20] Instead, an antiscalant addition is practiced which inhibits but cannot prevent $CaSO_4$ scaling, making it more difficult to control than other types. Table 1.1 lists several commonly dosed chemicals and antiscalants and their dosing concentrations.

1.1.2 Ion Exchange as a Pretreatment Method

Most methods for increasing the recovery of desalination processes focus on recovering water from the produced concentrate using forward osmosis to further concentrate the brine, or inducing precipitation of common scaling compounds as an intermediate treatment step.[24,29–38]

The removal of sulfate from the feedwater would prevent the formation of $CaSO_4$ during RO. Ion exchange resins may serve as a simple and effective method for the selective removal of SO_4^{2-} from background ions as demonstrated in

$$2\overline{\left(R_4N^+\right)Cl^-} + SO_4^{2-} \rightarrow \overline{\left(R_4N^+\right)_2 SO_4^{2-}} + 2Cl^- \tag{1.1}$$

In this case, sulfate is being selectively removed and replaced by chloride. Compared with sulfate salts, the solubility of chloride salts are orders of magnitude higher. Therefore, by passing the feed brackish water through a column of anion exchange resin preloaded in chloride form, the sulfate will be selectively removed and replaced by the more soluble chloride, and the RO process can be operated at higher recoveries without any threat to $CaSO_4$ scaling.

Ion exchange resins have a fixed capacity, and eventually all chloride will be exhausted and the column will need to be regenerated. During typical ion exchange regeneration, concentrated chloride brine would be passed through the column to displace sulfate in favor of chloride. In the place of a prepared regenerant solution, the concentrate stream itself may serve as a substitute regenerant, as it is concentrated chloride brine, resulting in a cyclic ion exchange–RO process that does not require any external regenerant.

1.1.3 HIX–RO PROCESS OVERVIEW

A flow chart of the proposed process is shown in Figure 1.1, and the individual steps of the process are as follows:

1. Influent feed solution is passed through a column of ion exchange resin in chloride form and sulfate is selectively removed by the following reaction:

$$2\overline{\left(R_4N^+\right)Cl^-} + SO_4^{2-} \rightarrow \overline{\left(R_4N^+\right)_2 SO_4^{2-}} + 2Cl^- \tag{1.2}$$

In this reaction, the overbar denotes the resin phase and R_4N^+ is the fixed functional group of the resin. For this step, the ion exchange column needs to be properly designed such that at the influent feedwater concentration, sulfate is preferred over chloride.

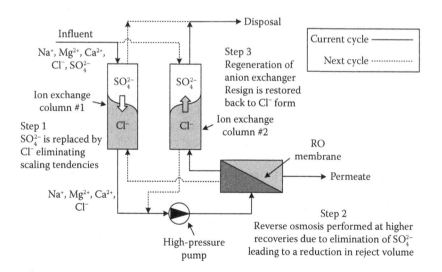

FIGURE 1.1 Flow chart of HIX–RO process.

2. After passing through the ion exchange column, the effluent should have little to no sulfate, and desalination by RO at higher recoveries is possible without any threat to $CaSO_4$ scaling.
3. The concentrate stream from the RO process, highly concentrated chloride brine, is used as a regenerant for the exhausted ion exchange and no external chemical input is needed. Sulfate is replaced by chloride through the following reaction:

$$\overline{\left(R_4N^+\right)_2 SO_4^{2-}} + 2Cl^- \rightarrow 2\overline{\left(R_4N^+\right)Cl^-} + SO_4^{2-} \tag{1.3}$$

Again, the ion exchange column must be designed properly to ensure that the concentration of the RO concentrate is high enough that sulfate is preferred over chloride. A combination of a properly designed ion exchange column with a high fraction of chloride in solution means that the regeneration process should be thermodynamically favorable.

Upon completion of desalination, a lower volume of reject stream is then discarded in the same manner as previously practiced at the RO treatment facility. No additional chemical input or change in concentrate management practices is necessary.

1.1.4 Previous Research on Ion Exchange–Assisted Desalination

Cyclic ion exchange processes have been studied in the past. The earliest examples include a French Patent from 1938 and a US Patent from 1946, both describing the use of ion exchange as a pretreatment method for boiler feed and using the waste stream from the blowdown as a regenerant.[39,40] Initial studies during 1950–1960 researched using cation exchange for softening seawater by selectively replacing calcium by sodium before use in boilers and using the blowdown to regenerate the resin.[41,42] Dow Chemical had also published work on their own research on seawater softening systems based on cyclic cation exchange desalination systems.[43] In the 1970s, researchers began trying to apply the same cyclic cation exchange/desalination systems to brackish water desalination.[44] In fact, early plans for the Yuma Desalination Plant in Arizona included the use of cyclic ion exchange–RO processes.[45] Researchers continued to focus on cation exchange pretreatment for calcium removal up until 2008.[46–52] No studies found before 2008 used anion exchange as a pretreatment method.

An early study performed in our lab researched the replacement of chloride with sulfate to reduce the osmotic pressure of the feed solution, allowing reduced energy requirements.[53] In 2012, researchers described the modeling and removal of divalent cations using ion exchange and a multistage RO system.[54] To date, no research on the use of mixed anion exchange resin columns has been found during literature review. Furthermore, besides the previous study in our lab, no research on cyclic anion exchange desalination systems was found in the open literature.[55]

1.1.5 CONTROL OF SULFATE REMOVAL BY MIXING OF ION EXCHANGE RESINS

The inherent success of the hybrid ion exchange–reverse osmosis (HIX–RO) process relies on the fact that resin selectivity toward sulfate or chloride is not absolute and can vary with total ionic strength. The relative preference for an ion exchange resin for one ion over another ion is the separation factor, α. Here, α is calculated by using Equation 1.4, where it takes into account the fraction of each species both on the resin, y, and in solution, x. The subscripts S and C indicate sulfate and chloride, respectively[56]

$$\alpha_{S/C} = \frac{y_S x_C}{x_S y_C} \tag{1.4}$$

Depending on the type of resin chosen, the selectivity of the resin toward sulfate will change. There are two main parameters that can be chosen for a given anion exchange resin: (i) the composition of the matrix and (ii) the functional group of the resin. The resin matrix is an insoluble cross-linked polymer that makes up the composition of the bead. The most commonly found resins are those made of polystyrene or polyacrylate. Given the same functional group, polyacrylic resins will remove more sulfate than polystyrene resins. Furthermore, for resins with amine functional groups, the sulfate preference follows the following sequence[57–59]:

primary > secondary > tertiary > quaternary

For example, the feedwater detailed in Table 1.2 is agricultural drainage water with high sulfate concentration from the San Joaquin Valley, California.[27]

At 80 meq/L feedwater concentration, Figure 1.2 shows theoretically generated isotherms for a strong base polystyrene resin at 80 meq/L influent concentration and 400 meq/L RO concentrate concentration. These curves were generated based on selectivity data from Clifford and Weber Jr.[57] Sulfate will be selectively removed for isotherm curves above the dashed line while chloride will be selectively removed for

TABLE 1.2

Composition of San Joaquin Valley Agricultural Drainage Water

Component	Value
TDS	5250 mg/L
Na^+	1150 mg/L
Mg^{2+}	60.7 mg/L
Ca^{2+}	555 mg/L
Cl^-	2010 mg/L
HCO_3^-	291 mg/L
SO_4^{2-}	1020 mg/L
pH	7.7
Conductivity	8.26 mS

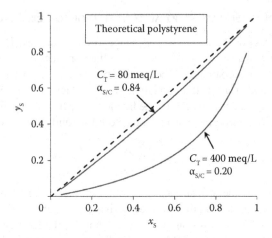

FIGURE 1.2 Theoretical polystyrene selectivity curve at 80 and 400 meq/L.

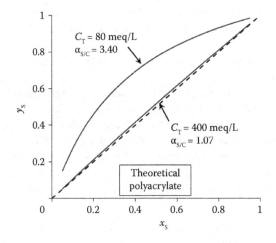

FIGURE 1.3 Theoretical polyacrylate selectivity curve at 80 and 400 meq/L.

curves below the diagonal. The axis labels x_S and y_S represent the fraction of sulfate in solution and on the resin, respectively, and $\alpha_{S/C}$ represents the selectivity coefficient of sulfate compared with chloride. When $\alpha_{S/C}$ is greater than one, sulfate is the preferred species, and when $\alpha_{S/C}$ is less than one, chloride is the preferred species. The polystyrene resin alone would be unsuitable for the given feedwater because of the low sulfate selectivity at feedwater concentration.

The same analysis can be performed for a polyacrylate resin, shown in Figure 1.3. Here, $\alpha_{S/C}$ is always greater than one, meaning that efficient regeneration of the resin will not occur.

In order for regeneration of the resin to be favorable, the resin must possess high chloride selectivity. In this case, the polystyrene resin has the desired selectivity

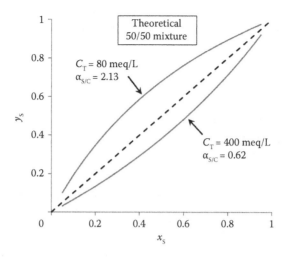

FIGURE 1.4 Theoretical 50/50 mixture of polystyrene and polyacrylic resins at 80 and 400 meq/L.

while the polyacrylic resin is still sulfate selective, and in order for the HIX–RO process to be sustainable, $\alpha_{S/C}$ must be greater than one at feedwater concentrations, and $\alpha_{S/C}$ should be less than one at RO concentrate concentrations. Neither the polystyrene resin nor the polyacrylic resin possesses this quality. However, if the two resins were mixed in a 1:1 ratio, as shown in Figure 1.4, the selectivity curves match up with the desired criteria: high sulfate selectivity at 80 meq/L (feedwater concentration) and high chloride affinity at 400 meq/L (RO reject concentration).

1.1.6 Reduction in $CaSO_4$ Scaling

The ultimate goal of HIX–RO is to reduce the influent sulfate concentration to prevent the formation of solid $CaSO_4$. In order to determine if $CaSO_4$ precipitation is thermodynamically favorable, the saturation index (SI) can be calculated using the following equation for a given a feedwater composition[20]:

$$\mathrm{SI} = \frac{\left\{ Ca^{2+} \right\}\left\{ SO_4^{2-} \right\}}{K_{sp}} \tag{1.5}$$

in which the curly brackets indicate species activity and K_{sp} is the solubility product for $CaSO_4$. The SI for $CaSO_4$ is plotted against the recovery of the RO process in Figure 1.5 for both the feedwater and if 80% of sulfate was removed and replaced by chloride. Both curves were generated using the water modeling software Stream Analyzer by OLI Systems.[60]

Based on the theoretical isotherm curves generated in Figure 1.4, a 50/50 mixture of strong base polyacrylic and strong base polystyrene resins should be able to ensure high sulfate removal for the feedwater composition given in Table 1.2. For a high sulfate removal, recovery of 80% is possible and the concentration of the RO reject

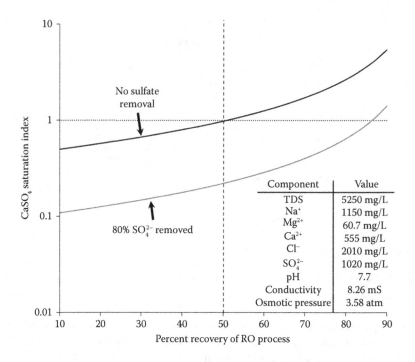

FIGURE 1.5 Variation in SI of $CaSO_4$ with recovery of desalination process.

brine is high enough to induce selectivity reversal, resulting in efficient regeneration of the ion exchange column. No additional chemical input would be required other than the concentrate from the RO process.

1.1.7 GOALS AND OBJECTIVES

1. Demonstrate that the HIX–RO process is able to reduce sulfate, operate at higher recoveries, and prevent scaling of $CaSO_4$. A properly designed HIX–RO system should be able to reduce the total concentration of sulfate so as to prevent the formation of $CaSO_4$ precipitate when operating the RO system at higher recoveries. Furthermore, this should be sustainable and not require any additional chemical input from the user; only the RO concentrate brine will be used as a regenerant.

2. Control resin selectivity through mixing of characteristically different resins. For a commercially available ion exchange resin that does not possess the correct $\alpha_{S/C}$ values, mixing two different resins will allow the user to directly control and tune $\alpha_{S/C}$ to the desired values.

3. Demonstrate that resin selectivity must be properly designed in order to achieve high sulfate removal. For systems where the ion exchange column has not been properly designed, little to no sulfate removal should be observed.

4. Quantify the potential for in-column precipitation of $CaSO_4$. During regeneration, because of the high concentration of sulfate, the water composition within the column will temporarily be supersaturated with $CaSO_4$ and there is a potential for precipitation and clogging. However, the kinetics of $CaSO_4$ precipitation are slow. It is, therefore, necessary to determine if in-column precipitation of $CaSO_4$ is a concern.

1.2 BACKGROUND ON ION EXCHANGE CHEMISTRY

The HIX–RO process is based on the selective removal of sulfate by a chloride-loaded strong base anion exchange resin as detailed in the following equation:

$$2\overline{(R_4N^+)Cl^-} + SO_4^{2-} \rightarrow \overline{(R_4N^+)_2\,SO_4^{2-}} + 2Cl^- \tag{1.6}$$

The equilibrium constant, K, for an ion exchange reaction is known as the selectivity coefficient, and for the reaction between sulfate and chloride, $K_{S/C}$ can be calculated by

$$K_{S/C} = \frac{\overline{\left[SO_4^{2-}\right]}\left[Cl^-\right]^2}{\left[SO_4^{2-}\right]\overline{\left[Cl^-\right]}^2} \tag{1.7}$$

If chloride and sulfate are the only anionic species present, then the fraction of sulfate in solution, x_S, and the total equivalent concentration, C_T, of anions in solution in eq/L are

$$x_S = \frac{2\left[SO_4^{2-}\right]}{\left[Cl^-\right]+2\left[SO_4^{2-}\right]} = \frac{2\left[SO_4^{2-}\right]}{C_T} \tag{1.8}$$

Similar calculations can be performed to calculate the fraction of each species on the resin, y_S:

$$y_S = \frac{2\overline{\left[SO_4^{2-}\right]}}{Q} \tag{1.9}$$

where Q is the total resin capacity in eq/g. For chloride, x_C and y_C are calculated in the same manner, and substituting Equations 1.8 and 1.9 into Equation 1.7 results in

$$K_{S/C} = \frac{y_S x_C^2}{x_S y_C^2}\frac{C_T}{Q} \tag{1.10}$$

From a purely theoretical point of view, $K_{S/C}$ is not a constant and may vary depending on experimental conditions such as shrinking/swelling of the resin, temperature changes, and so on. However, because of the small amount over which $K_{S/C}$ can change, it is a valid assumption that $K_{S/C}$ remains constant.[56,61,62] The units of $K_{S/C}$ are extremely important. While x and y are unitless, C_T and Q may be expressed in different units. Most commonly, C_T has units of [meq/L] and Q has units of [meq/g], making the overall units of $K_{S/C}$ [g/L]. From these terms, the separation factor, denoted by $\alpha_{S/C}$, may be calculated:

$$\alpha_{S/C} = \frac{y_S x_C}{x_S y_C} \tag{1.11}$$

When $\alpha_{S/C} > 1$, sulfate is the preferred species, and vice versa. The term $\alpha_{S/C}$ is significantly different from $K_{S/C}$; $\alpha_{S/C}$ is *not* a constant. The value of $\alpha_{S/C}$ will vary depending on several factors, most importantly solution concentration and composition. Changing the total solution concentration, C_T, has a more significant effect on $\alpha_{S/C}$ than changing the solution composition, that is, x_S or x_C.

For the HIX–RO system, C_T is not constant and it changes depending on the feedwater or RO concentrate composition. It becomes beneficial to determine $\alpha_{S/C}$ at various concentrations without having to run an isotherm at different C_T values.

Furthermore, if the resin chosen does not fall within the desired range of $\alpha_{S/C}$ values, it is possible to mix together characteristically different resins to achieve the desired range. Both of these methods are described in the following two sections.

1.2.1 DETERMINATION OF RESIN SEPARATION FACTOR FROM THE SELECTIVITY COEFFICIENT

At least one isotherm must be performed to determine $K_{S/C}$. Afterwards, $\alpha_{S/C}$ may be determined theoretically for any C_T. For a binary system where sulfate and chloride are the only anions present,

$$x_S + x_C = 1 \tag{1.12}$$

and

$$y_S + y_C = 1 \tag{1.13}$$

Equations 1.12 and 1.13 are rearranged and substituted into Equation 1.10 and rearranged to

$$\frac{y_S}{\left(1 - y_S\right)^2} = \frac{KQx_S}{\left(1 - x_S\right)^2 C_T} \tag{1.14}$$

and y_S can be determined using the quadratic equation

$$y_S = \frac{\dfrac{2A+1}{A} - \sqrt{\left(\dfrac{2A+1}{A}\right)^2 - 4}}{2} \tag{1.15}$$

in which

$$A = \frac{KQx_S}{\left(1 - x_S\right)^2 C_T} \tag{1.16}$$

Only the negative root of the solution is a valid solution as the positive root gives a value of y_S greater than 1. Simplifying Equation 1.15 results in

$$y_S = \frac{2A + 1 - \sqrt{4A + 1}}{2A} \tag{1.17}$$

And $\alpha_{S/C}$ may be determined using Equation 1.11. This solution is valid assuming $K_{S/C}$ is constant at a given x_S and C_T.

1.2.2 Theoretical Prediction of Resin Separation Factor for Two Mixed Resins

For a mixture of two resins, the separation factor, $\alpha_{S/C}$, for each individual resin remains the same, but the bulk separation factor, $\alpha_{S/C}^*$, is different and describes the entire resin mixture. For a mixture of two different resins A and B with masses m_A and m_B, the ratio of resin A, Φ_A, is

$$\Phi_A = \frac{m_A}{m_A + m_B} \tag{1.18}$$

and similarly for resin B

$$\Phi_B = \frac{m_B}{m_A + m_B} = 1 - \Phi_A \tag{1.19}$$

The total ion exchange capacity of the system is therefore

$$Q^* = Q_A \Phi_A + Q_B (1 - \Phi_A) \tag{1.20}$$

Calculation of $\alpha_{S/C}^*$ is similar to Equation 1.11:

$$\alpha_{S/C}^* = \frac{y_S^* x_C}{x_S y_C^*} \tag{1.21}$$

in which

$$y_S^* = \frac{y_S^A Q_A \Phi_A + y_S^B Q_B (1 - \Phi_A)}{Q^*} \tag{1.22}$$

and y_S for either resin may be calculated using Equation 1.17.

Using this method, it is possible to generate a range of mixing ratios over which the HIX–RO process will be favorable. Using the same selectivity data used to create Figures 1.2 and 1.3, Figure 1.6 illustrates the change in $\alpha_{S/C}^*$ with different mixings for the San Joaquin Valley water (composition given in Table 1.2).

Any of the mixing ratios within the shaded area will give the desired range of $\alpha_{S/Cl}^*$ values: greater than one at feedwater concentration (80 meq/L) and less than one at RO reject concentration (400 meq/L).

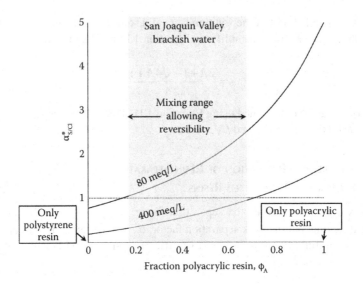

FIGURE 1.6   Variation in theoretical $\alpha^*_{S/Cl}$ with changing mixing ratio.

1.3   EXPERIMENTAL METHODS

Throughout this study, three different types of ion exchange resins were used as shown in Table 1.3.

1.3.1   ION EXCHANGE RESINS

All resins were strong base but varied in resin matrix (polystyrene or polyacrylic) or functional group (trimethylamine or triethylamine).

1.3.1.1   Sample Analysis

Chloride was analyzed by the argentometric titration method. Samples were diluted to a volume of 100 mL, placed in a 250 mL Erlenmeyer flask, and the pH was adjusted to between 7 and 10 using 2% NaOH. One milliliter of 5% K_2CrO_4 indicator was added. The sample was placed on a stir plate and titrated against standardized 0.0141 N $AgNO_3$. Chloride precipitates with Ag^+ to form solid AgCl. Once all the Cl^- has been precipitated, Ag^+ will form the dark red precipitate Ag_2CrO_4 and the solution will change color from yellow to pink.[63]

Analysis of the sulfate was performed using a commercially available sulfate testing kit available from the Hach Company (SulfaVer 4 Method #8051). An aliquot of 10 mL is placed in a glass sample cell and one pillow of powdered $BaCl_2$ is added. The cell is then swirled to dissolve $BaCl_2$, and any sulfate present will precipitate as $BaSO_4$. The sample is left to react for 5 min and then analyzed using a Hach Spectrophotometer (Model DR 7000) in which the absorbance of the sample is directly correlated to mg/L as SO_4^{2-}.[64]

TABLE 1.3

Properties of Anion Exchange Resins Used in This Study

Manufacturer	Purolite, Inc.	Purolite, Inc.	Rohm and Haas Co.
Trade Name	A400	A850	IRA-900
Type	Strong base	Strong base	Strong base
Matrix	Polystyrene	Polyacrylic	Polystyrene
Functional Group	Quaternary trimethylamine	Quaternary trimethylamine	Quaternary triethylamine
Structure			

1.3.1.2 Resin Capacity Measurement

Anion exchange resins were added in chloride form by packing a glass column with a known mass of air-dried resin and passing a dilute sodium chloride solution through the column until the influent and effluent solution had the same concentration of chloride. The columns were then washed with deionized (DI) water. Next, a dilute sulfate solution was prepared and passed through the conditioned resin. The effluent was collected in a container until the influent and effluent sulfate concentrations were equivalent. The total volume of collected effluent was measured, and the resin capacity was calculated by measuring the total concentration of sulfate and chloride. The total mass of chloride in the effluent solution was then assumed to be the total mass of chloride present on the column, and therefore the resin capacity.

1.3.1.3 Batch Sulfate/Chloride Isotherms

Resins were first conditioned by packing in a glass column and passing a dilute sodium chloride solution until the influent and effluent chloride concentrations were equal. The column was then washed with DI water. The resins were then removed from the glass column and placed on the lab bench to air dry at room temperature for at least 48 h.

Next, varying masses of air-dried ion exchange resin were placed in plastic bottles. A stock solution of sulfate at the desired isotherm concentration was prepared, and equal volumes of solution were added to each bottle containing a known mass of resin. The bottles were then capped, sealed with Parafilm, and placed on a rotary shaker for at least 24 h. The solution was decanted from the resin and the composition was analyzed for chloride and sulfate.

1.3.1.4   Column Sulfate/Chloride Isotherms

A known mass of air-dried ion exchange resin was packed in a glass column, and a dilute sodium chloride solution was passed through the column at a constant flow rate using a ceramic peristaltic pump until the influent and effluent chloride concentrations were equal. The column was washed with DI water, and a prepared solution containing both chloride and sulfate was passed. The effluent solution was collected in glass test tubes using an Eldex Universal Fractional Collector, and the column was run until the effluent chloride and sulfate concentrations matched the influent concentration.

1.3.1.5   Scanning Electron Microscopy

Cross-sectional analysis of ion exchange resin was performed using a Philips XL-30 Environmental Scanning Electron Microscope. Samples were prepared by slicing individual ion exchange beads with razor blades that had been immersed in liquid nitrogen. The bead halves were mounted onto pegs using double-sided carbon tape and sputter coated with iridium using an Electron Microscopy Sciences high vacuum sputter coater (Model EMS575X).[65]

1.3.2   HIX–RO RUNS

For each HIX–RO run, 20 L of influent solution was prepared and passed downflow through 1 L of ion exchange resin. The effluent solution was collected and subjected to RO. Finally, the concentrate stream from RO was passed upflow through the ion exchange column. The process of ion exchange, RO, and ion exchange regeneration constitutes one "cycle" and a group of cycles was considered to be a "run." Before beginning a run, the resin was first conditioned by passing a dilute chloride solution until the influent and effluent chloride concentrations were equal and washed with DI water. In between cycles, the resin bed was not washed or disturbed in any way. Any remaining solution from a previous cycle remained inside the column at the start of the next cycle.

A custom-made ion exchange column was used during all HIX–RO cycles. The main body of the column was constructed from clear PVC with screw-on PVC end caps. The inner diameter was 5 cm and the total length of the column was 68 cm. Glass wool was packed into the top and bottom to prevent any loss of ion exchange resin during use. The solution was fed using a variable speed peristaltic pump.

The influent solution was stored in a polyethylene tank and fed using a stainless steel piston pump (Cat Pumps, Model 2SF35SEEL) powered by a 1.5-hp electric motor. A filter (GE SmartWater GXWH20F) was placed before the membrane to prevent damage from any particulate matter in the feedwater. RO was performed using a Dow Filmtec SW30-2540 Spirally Wound Reverse Osmosis Membrane with an effective membrane area of 2.8 m². All tubing was made of 316 stainless steel to avoid corrosion. A cooling coil was immersed in the feed solution and tap water was fed through to maintain temperature at 20°C–25°C.

The influent solution was prepared based on the composition given in Table 1.2. The feedwater solution was modified such that any equivalent concentration of bicarbonate was converted to chloride under the assumption that hydrochloric acid had

TABLE 1.4

Feedwater Composition

for All HIX–RO Runs

Ion	Concentration (meq/L)
Na^+	75
Mg^{2+}	5
Cl^-	60
SO_4^{2-}	20

been dosed to eliminate the threat to carbonate scaling. In addition, all calcium was converted to an equivalent amount of sodium to ensure that no scaling occurs during the cycles. The final feedwater composition used for all runs is detailed in Table 1.4.

Two runs of HIX–RO were performed: one using a 50/50 mixture of strong base polystyrene and polyacrylic resins, and a second run containing only polystyrene but with trimethylamine functional groups.

1.3.3   Measuring $CaSO_4$ Precipitation Kinetics

The induction time for $CaSO_4$ precipitation was measured by setting up a time lapse experiment. Two stock solutions of 0.03 M $CaCl_2$ and 0.03 M Na_2SO_4 were prepared using ACS grade chemicals. Equal volumes of stock solution were then mixed with varying ratios of DI water in 20 × 170 mm test tubes to form supersaturated solutions of $CaSO_4$. The exact SI values were calculated using an OLI Stream Analyzer.[60] The test tubes were covered in Parafilm and vigorously shaken until the solution was well mixed. The test tubes were then placed in a test tube rack in front of a black background. A digital video camera with a time lapse function was programmed to take a picture every 5 s and recorded images of the test tubes for 27 h. After recording, the footage was analyzed frame by frame to determine the exact time when the first visible crystal of $CaSO_4$ precipitated.

1.3.4   In-Column $CaSO_4$ Precipitation

The potential for in-column precipitation of $CaSO_4$ was studied using a small-scale column setup. An 11 mm glass column was filled with a 50/50 mixture, by mass, of polystyrene anion exchange resin and polyacrylic anion exchange resin. A dilute solution of NaCl was passed through the column to ensure the resin was in chloride form. The column was then washed with DI water until the conductivity of the effluent matched that of the influent.

A synthetic influent solution was prepared that simulated the actual feedwater composition given in Table 1.2. The regenerant solution, synthetic RO concentrate, was a simulated RO concentrate that had the synthetic influent solution previously subjected to ion exchange and RO at 80% recovery. Table 1.5 gives the exact composition for each solution.

TABLE 1.5

Composition of Synthetic Feedwater and Regenerant Solution

	Synthetic Feedwater (meq/L)	Synthetic Regenerant (meq/L)
Na^+	50	250
Mg^{2+}	5	25
Ca^{2+}	25	125
Cl^-	60	400
SO_4^{2-}	20	–

During one cycle, 20 bed volumes (BVs) of synthetic influent solution were passed downflow through the column and collected for analysis. Next, 4 BVs of synthetic RO concentrate solution were passed upflow through the column and collected using a fractional collector. The empty-bed contact time for the column was purposely kept as low as possible to avoid any possible in-column precipitation. Immediately after passing the regenerant, another 20 BV of influent solution was passed to ensure that there was never an extended period where the column could be in a supersaturated state.

The process of passing synthetic influent followed by passing synthetic RO concentrate constituted one cycle. For the first three cycles, after passing the regenerant solution, the collected samples were covered in Parafilm and left to precipitate. After at least 24 h, samples were then taken and measured for calcium and sulfate to determine the post-precipitation concentration of calcium and sulfate. During the fourth regeneration cycle, samples from the effluent of the column were taken and immediately diluted 1:100 and analyzed for calcium and sulfate to determine the pre-precipitation concentration of calcium and sulfate in solution.

1.4 RESULTS/DISCUSSION

1.4.1 CLASSIFICATION OF RESIN PROPERTIES AFFECTING SULFATE SELECTIVITY

There are several properties of the ion exchange resin that may affect the overall resin sulfate selectivity:

1. Composition of the polymer matrix
2. Size of the functional group
3. Basicity of the functional group

Matrix type refers to the base polymer that has been functionalized to give ion exchange capabilities. The two most commonly used polymers are polyacrylate and polystyrene. Variations in the basicity and size of the functional group also affect selectivity. Weak base resins selectivity remove divalent ions more than resins with strong base functional groups.[58] The size of the functional group also affects selectivity. Steric hindrance between an ion and the functional group will change the overall selectivity. Replacing a quaternary ammonium functional group $(R-N^+(CH_3)_3)$ with a triethylamine group $(R-N^+(CH_2CH_3)_3)$ hinders the ability of divalent ions like sulfate to interact with the functional group, resulting in lowered sulfate selectivity.[66]

Comparing resin matrix effects only, the polyacrylic resin has a higher affinity toward sulfate than a polystyrene resin with the same functional groups. This effect can be attributed to the fact that the polyacrylic resin has a higher capacity than the polystyrene resin, implying that for a given ion exchange bead, there are more active sites for the sulfate to interact with, making adsorption easier.

The effects of changing the basicity of the functional group are shown in Figure 1.7. The isotherms for the polyamine and tertiary amine resins are garnered from selectivity data in the open literature.[57]

Resins with weaker base functional groups exhibit higher affinities toward sulfate than those with strong base functional groups. In other words, the order of sulfate selectivity follows:

$$polyamine > tertiary > quaternary$$

In total, resins with a polyacrylic matrix or weak base functional groups have a higher affinity toward sulfate than resins with a polystyrene matrix or strong base functional group.

The large variation in types of anion exchange resin and large variation in selectivity mean that it should always be possible to achieve the desired $\alpha^*_{S/C}$ values through either judicious choosing of resin type or mixing different resins.

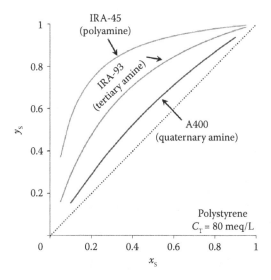

FIGURE 1.7   Effect of changing basicity of functional group.

1.4.2 EXPERIMENTAL MEASUREMENT OF INDIVIDUAL RESIN SEPARATION FACTOR

Six batch studies were performed for all three resin types at 80 meq/L total ion concentration. The resulting isotherm curves generated from these batch tests are shown in Figure 1.8.

The separation factor, $\alpha_{S/Cl}$, was determined by using Equation 1.11 from the measured data and using Equation 1.10, $K_{S/C}$ values were also determined. A list of the pure resin parameters is provided in Table 1.6.

1.4.3 EXPERIMENTAL MEASUREMENT OF MIXED RESIN SEPARATION FACTORS

Both batch and column isotherms were performed to determine mixed resin selectivity. From theoretical predictions, a 50/50 mixture of polystyrene and polyacrylic resins would result in the highest efficiency. Therefore, isotherms at both 80 meq/L and 400 meq/L were carried out for the 50/50 resin mixture and are shown in Figure 1.9.

In addition to the 50/50 mixture, isotherms were performed on two other mixing ratios at 80 meq/L: 25:75 and 75:25. A summary of the results from pure and mixed isotherms is shown in Figure 1.10 along with the theoretical $\alpha_{S/Cl}$.

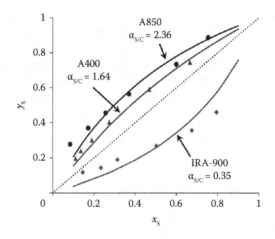

FIGURE 1.8 Batch testing results for strong base polystyrene, strong base polyacrylic resin, and strong base polystyrene with triethylamine functional group at 80 meq/L.

TABLE 1.6
Measured Pure Resin Parameters

	A400	A850	IRA-900
Q (meq/g dry resin)	1.8	2.2	3.6[67]
$\alpha_{S/C}$ at 80 meq/L	1.68	2.36	0.35
$K_{S/C}$ (g/L)	85.1	114.9	6.47

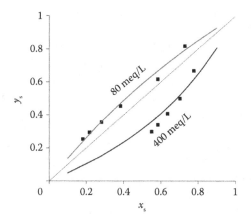

FIGURE 1.9 Batch isotherms for a 50/50 mixture of polystyrene and polyacrylic resins at 80 and 400 meq/L.

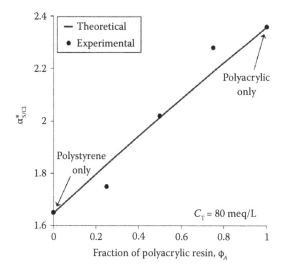

FIGURE 1.10 Variation in $\alpha^*_{S/Cl}$ with changing fraction of polyacrylic resin at 80 meq/L.

It is clear from Figure 1.10 that theoretical predictions of $\alpha^*_{S/Cl}$ based on the pure resin isotherm data align well with experimental data.

1.4.4 RESULTS FROM HIX–RO RUNS

Since the concern is the prevention of calcium sulfate scaling, data regarding the quality of the RO permeate are not presented. For both runs, the conductivity of the permeate never exceeded 1000 μS. This indicates that the integrity of the RO membrane was not compromised during any HIX–RO runs, and that little to no ions were lost because of permeation through the membrane.

1.4.4.1 Mixed Bed Polystyrene and Polyacrylic Resins

For this run, the ion exchange bed was a 50/50 mixture of polystyrene and poly-acrylic resins: a properly designed HIX–RO system. The concentration of sulfate at the exit of the ion exchange column (RO feed) is shown in Figure 1.11.

For 10 successive cycles, the sulfate concentration at the exit of the ion exchange column was reduced by 90% or greater: a significant reduction. However, the issue is not the presence of sulfate but its potential for forming $CaSO_4$ during RO. Since no Ca^{2+} was present in the synthetic feedwater solution, theoretical calculations must be performed to determine the concentration of Ca^{2+} that would be present in the con-centrate. Using the actual feedwater Ca^{2+} concentration, the measured recovery for each cycle, and assuming no loss through the RO membrane, the theoretical concen-tration of Ca^{2+} can be calculated. By combining the theoretical Ca^{2+} concentration with the measured SO_4^{2-} values, the SI for $CaSO_4$ was determined using OLI and is shown in Figure 1.12; note that the y-axis is plotted log-scale.[60]

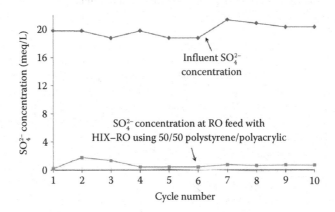

FIGURE 1.11   Concentration of sulfate at RO feed.

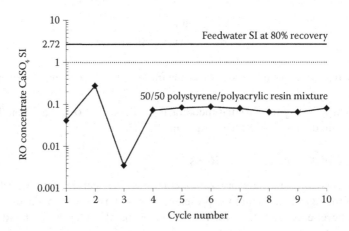

FIGURE 1.12   Calculated SI values of $CaSO_4$ in RO concentrate.

This method of back-calculating the concentration of Ca^{2+} in order to determine $CaSO_4$ SI values in the RO concentrate was performed for both runs. For all 10 cycles, the SI never exceeded one and there was never a threat of $CaSO_4$ scaling.

The sustainability of the HIX–RO process is demonstrated by performing a mass balance on sulfate for the entire HIX–RO system. For each cycle, the amount of sulfate entering (from the influent) and exiting the system (from the waste ion exchange regenerant solution) was determined and is shown in Figure 1.13.

Results show that the mass of sulfate entering and exiting the system during each cycle was approximately equal. The mass balance over the entire 10 cycles was off by 0.35 meq sulfate, which is negligible considering that the total mass of sulfate entering the system over the 10 cycles was approximately 4000 meq. For chloride, a mass balance on the ion exchange column was performed and is shown in Figure 1.14.

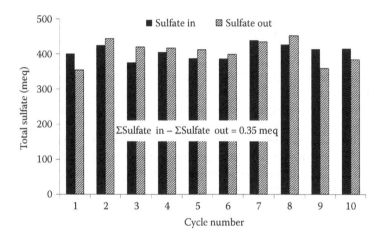

FIGURE 1.13 Mass balance on sulfate-entering and -exiting system.

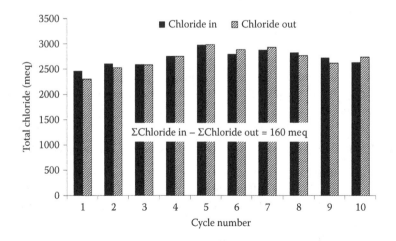

FIGURE 1.14 Mass balance on chloride-entering and -exiting system.

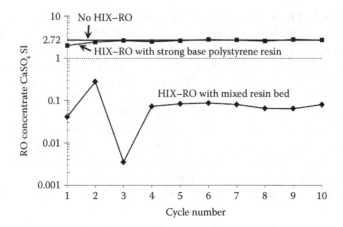

FIGURE 1.15   Calculated SI values for $CaSO_4$ in RO concentrate.

For each cycle, the mass of chloride entering and exiting the IX column was approximately equal, and over the entire 10 cycles, the mass balance on chloride was off by 160 meq. Again, this difference is negligible considering that the total mass of Cl^- entering the system was about 12,000 meq.

While the use of a mixed bed anion exchange column to reduce sulfate was successful, it is equally important to demonstrate that not just any type of anion exchange resin can be chosen. An improperly designed ion exchange column would result in incomplete regeneration or lower sulfate removal. To this end, another HIX–RO run with a resin low in sulfate affinity was performed.

1.4.4.2   Pure Polystyrene with Triethylamine Functional Groups

This HIX–RO run was performed with only a polystyrene resin with triethylamine functional groups. Using the isotherm data, it is predicted that $\alpha_{S/Cl}$ for IRA-900 is 0.35 at 80 meq/L, so very little removal was expected. Indeed, low sulfate removal was observed and as a result the SI, plotted log-scale in Figure 1.15, was not reduced enough to prevent $CaSO_4$ precipitation.

Compared to the mixed resin bed, the polystyrene resin with triethylamine functional groups was unable to reduce sulfate enough to prevent the formation of $CaSO_4$, demonstrating that if the resin selectivity has not been properly tuned, no sulfate removal will occur.

1.4.5   Characterization of Potential for In-Column Precipitation of $CaSO_4$

During regeneration of the ion exchange column, the effluent from the column is a highly concentrated brine of sulfate and the formation of $CaSO_4$ may be a problem. Precipitation of $CaSO_4$ inside the column could cause clogging of the ion exchange bed or, worse, inhibition of portions of the resin. This would result in not only higher head losses through the column but also reduced efficiency from the reduction in available ion exchange sites.

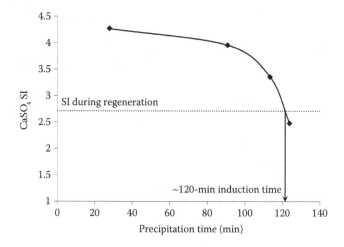

FIGURE 1.16 Time for visible precipitation of $CaSO_4$ with varying SI values.

1.4.5.1 Measurement of $CaSO_4$ Precipitation Kinetics

The precipitation of $CaSO_4$ is not instantaneous. The rate is dependent on several factors, including: induction time, presence of seed crystals, and how well mixed the bulk solution is.[37] It is therefore necessary to determine how quickly $CaSO_4$ will precipitate relative to a typical bed contact time.

Different volumes of stock solutions of $CaCl_2$ and Na_2SO_4 were added to large test tubes and mixed, creating supersaturated solutions of $CaSO_4$ with varying SI values. A video camera was set up to record when precipitation occurs for each SI value and the results are plotted in Figure 1.16. From the fitted line, the time for $CaSO_4$ to precipitate with an SI value of 2.72 (the SI value of the regenerant) is approximately 120 min.

1.4.5.2 Small-Scale In-Column Precipitation Study

A study on the potential for in-column precipitation of $CaSO_4$ was performed. Synthetic influent and regenerant solutions were passed through a 50/50 mixture of polystyrene and polyacrylic anion exchange resins. During regeneration, the empty-bed contact time was 10 min, which is significantly less than the predicted time for $CaSO_4$ precipitation.

Three consecutive cycles were run and no precipitate was observed within the mixed bed anion exchange column. However, after 120 min, visible precipitates were formed in the tubes of the sample collector, as shown in Figure 1.17.

For more quantitative information, samples were collected from the effluent during the third cycle. Samples were collected during regeneration and small aliquots of effluent regenerant were immediately diluted to prevent precipitation. Calcium and sulfate were analyzed in the diluted samples and after 24 h, and SI values were calculated using OLI.[60] The calculated SI values are plotted log-scale in Figure 1.18 against BVs of reject regenerant passed for the two sets of samples. While the effluent solution was, in fact, supersaturated with $CaSO_4$, no visible precipitation occurred inside the column.

FIGURE 1.17 Formation of $CaSO_4$ precipitates after 120 min.

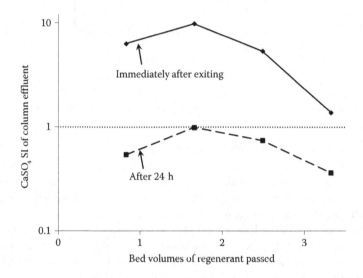

FIGURE 1.18 $CaSO_4$ SI values immediately after exiting and after 24 h.

In addition to visible inspection, samples of ion exchange resin were extracted from the column and analyzed using SEM-EDX (energy dispersive X-ray spectroscopy) to determine if any calcium was present inside the bead. Before analysis, the beads were washed with DI water and left to air dry for 24 h. Individual beads were then sliced using a razor blade dipped in liquid nitrogen and analyzed. Results from analyses of seven different ion exchange resin halves showed that none of the resins analyzed had any trace of calcium. A representative EDX spectrum is shown in Figure 1.19.

The composition of the bead contained carbon, nitrogen, chlorine, oxygen, and sulfur but no calcium. These elements are expected for a strong base anion exchange resin. Carbon is present in the resin matrix, nitrogen in the quaternary ammonium functional group, and chloride, sulfur, and oxygen would be any chloride or sulfate ions occupying the exchange sites.

1.4.5.3 Explanation for Lack of In-Column Precipitation

There are two reasons why no $CaSO_4$ precipitated inside of the ion exchange column. The first is that the bed contact time was kept at only 9.3 min, which is

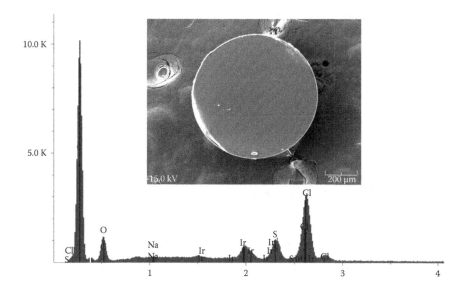

FIGURE 1.19   Representative EDX spectrum of resin bead.

significantly shorter than the kinetics of $CaSO_4$ formation. Even though the SI was greater than one for the first three BVs, precipitation would take more than 120 min (from Figure 1.16), but the bed contact time was significantly less. Similar results from previous research have demonstrated that if an ion exchange reaction results in thermodynamically favorable precipitation, no precipitation may occur; this concept is known as ion exchange induced supersaturation.[68]

The second reason for the lack of in-column $CaSO_4$ formation is because of Donnan membrane effects. The large number of positively charged sites on the ion exchange resin makes the transfer of Ca^{2+} to inside the bead highly unfavorable. For example, an anion exchange resin with 2 meq/g capacity can theoretically possess up to 1.20×10^{21} positive sites per gram of resin.

Estimation of the concentration of Ca^{2+} inside the resin can be performed by assuming that only Ca^{2+} and Cl^- are present in solution, and the resin is in chloride form. For these assumptions, the following equation is valid:

$$\left[Ca^{2+} \right]_i \left[Cl^- \right]_i^2 = \left[Ca^{2+} \right]_o \left[Cl^- \right]_o^2 \tag{1.23}$$

in which the subscripts "i" and "o" are the concentration of the ion inside and outside of the resin, respectively. If the total capacity of the resin is 2.2 M and x mol/L of Ca^{2+} enter the resin, then the amount of Cl^- inside the resin is therefore $2 + 2x$. If the regenerant concentration is 400 meq/L, then Cl^- will be 0.4 M and Ca^{2+} will be 0.2 M. Subbing these values into Equation 1.23 results in

$$(x)(2.2 + 2x)^2 = (0.2 \text{ M})(0.4 \text{ M})^2 \tag{1.24}$$

Solving for x gives an inter-resin Ca^{2+} concentration of 6.53×10^{-3} M or 13.1 meq/L, which is only ~3% of the total calcium in the regenerant.

Because of the inability of calcium to enter inside the resin combined with slow reaction kinetics, precipitation with sulfate inside is highly unfavorable.

1.5 CONCLUSIONS

Important conclusions that can be drawn from this study are as follows:

1. Resin selectivity is dependent on the resin type and functional group. A resin with weak base functional groups and/or a polyacrylic matrix will possess higher sulfate selectivity than one with strong base functional groups or a polystyrene matrix. In addition, a larger functional group, like triethylamine, will have lower sulfate selectivity than a smaller functional group, like trimethylamine.
2. Resin mixing allows precise control over the range of $\alpha_{S/C}$ values. Theoretical calculations of sulfate selectivity align well with experimentally measured selectivity values, indicating that it is possible to predict selectivity values for any resin mixing ratio as demonstrated in Figures 1.6 and 1.10. This is critical for the HIX–RO process where an incorrectly designed system will not effectively remove sulfate.
3. The large variations in type of resin and mixing mean that the HIX–RO process is not just limited to one feedwater. No matter what the sulfate concentration or C_T, there should always be some type of resin or resin mixture that will result in the desired range of $\alpha_{S/C}$ values so the resin can be precisely tuned to any feedwater. If a commercially available resin does not fit within the desired parameters, a mixture of two resins will allow the user to achieve the desired $\alpha_{S/C}$ values.
4. A properly designed HIX–RO system can achieve higher recoveries while keeping the $CaSO_4$ SI below 1. Results show that the supersaturation of $CaSO_4$ was prevented, higher than normal recoveries were achieved, and the process was sustainable. During cycles 5–10 for the 50/50 resin mixture, the effluent sulfate concentration was consistent, and the mass balance showed that there was no loss or accumulation of sulfate within the system further, demonstrating that a steady state had been reached.
5. Precipitation of $CaSO_4$ inside the column or resin bead was not observed. Due to the slow kinetics of $CaSO_4$ precipitation, the short empty-bed contact time of the ion exchange column, and the quick elution of sulfate from the column all prevent the formation of $CaSO_4$ precipitates inside the pore space of the column. Furthermore, Donnan membrane effects from the fixed functional charges on the anion exchange resin prevent Ca^{2+} from entering inside the resin and causing $CaSO_4$ precipitation. All these factors combined result in the lack of in-column or intra-resin precipitation.

ACKNOWLEDGMENTS

This research received funding from the National Science Foundation Accelerating Innovation Research (NSF-AIR 1311758) Grant and the Pennsylvania Infrastructure Technology Alliance (PITA).

REFERENCES

1. United States Environmental Protection Agency. Fiscal Year 2011: Drinking and Ground Water Statistics. http://water.epa.gov/scitech/datait/databases/drink/sdwisfed/upload/epa816r13003.pdf (accessed September 20, 2014).
2. Averyt, K., Meldrum, J., Caldwell, P., Sun, G., McNulty, S., Huber-Lee, A., Madden, N. Sectoral contributions to surface water stress in the coterminous United States. *Environ. Res. Lett.* **2013**, *8*, 035046.
3. Scanlon, B. R., Faunt, C. C., Longuevergne, L., Reedy, R. C., Alley, W. M., Mcguire, V. L., Mcmahon, P. B. Groundwater depletion and sustainability of irrigation in the US High Plains and Central Valley. *Proc. Natl. Acad. Sci. U. S. A.* **2012**, *109* (24), 9320–9325.
4. Ye, L., Grimm, N. B. Modelling potential impacts of climate change on water and nitrate export from a mid-sized, semiarid watershed in the US Southwest. *Clim. Change.* **2013**, *120*, 419–431.
5. MacDonald, G. M. Severe and sustained drought in Southern California and the West: Present conditions and insights from the past on causes and impacts. *Quat. Int.* **2007**, *173–174*, 87–100.
6. MacDonald, G. M. Climate change and water in Southwestern North America special feature: Water, climate change, and sustainability in the Southwest. *Proc. Natl. Acad. Sci. U. S. A.* **2010**, *107* (50), 21256–21262.
7. Sabo, J. L., Sinha, T., Bowling, L. C., Schoups, G. H. W., Wallender, W. W., Campana, M. E., Cherkauer, K. A., Fuller, P. L., Graf, W. L., Hopmans, J. W., et al. Reclaiming freshwater sustainability in the Cadillac Desert. *Proc. Natl. Acad. Sci. U. S. A.* **2010**, *107* (50), 21263–21270.
8. Gober, P., Kirkwood, C. W. Vulnerability assessment of climate-induced water shortage in Phoenix. *Proc. Natl. Acad. Sci. U. S. A.* **2010**, *107* (50), 21295–21299.
9. Reynolds, T. D., Richards, P. A. *Unit Operations and Processes in Environmental Engineering*, 2nd ed., PWS Publishing Company, Boston, MA, 1996.
10. National Research Council. *Desalination: A National Perspective*, The National Academies Press, Washington, DC, 2008.
11. Nicot, J., Walden, S., Greenlee, L., Els, J. A. Desalination Database for Texas. 2005. http://www.beg.utexas.edu/environqlty/desalination/Final%20Report_R1_1.pdf (accessed September 20, 2014).
12. Mickley, M. C. *Membrane Concentrate Disposal: Practices and Regulations. Desalination and Water Purification Research and Development Program.* Report No. 69, U.S. Department of the Interior, Denver, CO, 2001.
13. Abdul-Wahab, S. A., Al-Weshahi, M. A. Brine management: Substituting chlorine with on-site produced sodium hypochlorite for environmentally improved desalination processes. *Water Resour. Manag.* **2009**, *23*, 2437–2454.
14. Mickley, M. C. *Treatment of Concentrate. Desalination and Water Purification Research and Development Program.* Report No. 155, US Department of the Interior Bureau of Reclamation, Denver, CO, 2009.
15. Younos, T. Environmental issues of desalination. *J. Contemp. Water Res. Educ.* **2005**, *132*, 11–18.

16. Hasson, D., Drak, A., Semiat, R. Inception of CaSO$_4$ scaling on RO membranes at various water recovery levels. *Desalination*. **2001**, *139*, 73–81.

17. Le Gouellac, Y. A., Elimelech, M. Calcium sulfate (gypsum) scaling in nanofiltration of agricultural drainage water. *J. Membr. Sci.* **2002**, *205*, 279–291.

18. Shih, W., Rahardianto, A., Lee, R., Cohen, Y. Morphometric characterization of calcium sulfate dehydrate (gypsum) scale on reverse osmosis membranes. *J. Membr. Sci.* **2005**, *252* (1–2), 253–263.

19. Al-Amoudi, A., Lovitt, R. W. Fouling strategies and the cleaning system of NF membranes and factors affecting cleaning efficiency. *J. Membr. Sci.* **2007**, *303*, 4–28.

20. Gabelich, C. J., Williams, M. D., Rahardianto, A., Franklin, J. C., Cohen, Y. High-recovery reverse osmosis desalination using intermediate chemical demineralization. *J. Membr. Sci.* **2007**, *301* (1–2), 131–141.

21. Tzotzi, C., Pahiadaki, T., Yiantsios, S. G., Karabelas, A. J., Andritsos, N. A study of CaCO$_3$ scale formation and inhibition in RO and NF membrane processes. *J. Membr. Sci.* **2007**, *296*, 171–184.

22. Lyster, E., Au, J., Rallo, R., Giralt, F., Cohen, Y. Coupled 3-D hydrodynamics and mass transfer analysis of mineral scaling-induced flux decline in a laboratory plate-and-frame reverse osmosis membrane module. *J. Membr. Sci.* **2009**, *339*, 39–48.

23. Alhseinat, E. A., Sheikholeslami, R. A reliable approach for barite, celestite and gypsum scaling propensity prediction during reverse osmosis treatment for produced water. In *International Proceedings of Chemical, Biological, and Environmental Engineering, vol. 21, 2011 International Conference on Environment and BioScience*, IACIST Press, Singapore, 2011, pp. 52–57.

24. Antony, A., Low, J. H., Gray, S., Childress, A. E., Le-Clech, P., Leslie, G. Scale formation and control in high pressure membrane water treatment systems: A review. *J. Membr. Sci.* **2011**, *383*, 1–16.

25. Van Driessche, E. S., Benning, L. G., Rodriguez-Blanco, J. D., Ossorio, M., Bots, P., García-Ruiz, J. M. The role and implications of bassanite as a stable precursor phase to gypsum precipitation. *Science*. **2012**, *336*, 69–72.

26. Waly, T., Kennedy, M. D., Witkamp, G., Amy, G., Schippers, J. C. The role of inorganic ions in the calcium carbonate scaling of seawater reverse osmosis systems. *Desalination*. **2012**, *284*, 279–287.

27. Rahardianto, A. Diagnostic characterization of gypsum scale formation and control in RO membrane desalination of brackish water. *J. Membr. Sci.* **2006**, *279*, 655–668.

28. Florida Department of Environmental Protection. *Desalination in Florida: Technology, Implementation, and Environmental Issues*. 2010. http://www.dep.state.fl.us/water/docs/desalination-in-florida-report.pdf (accessed October 10, 2014).

29. Tang, W., Ng, H. Y. Concentration of brine by forward osmosis: Performance and influence of membrane structure. *Desalination*. **2008**, *224*, 143–153.

30. Martinetti, C. R., Childress, A. E., Cath, T. Y. High recovery of concentrated RO brines using forward osmosis and membrane distillation. *J. Membr. Sci.* **2009**, *331*, 31–39.

31. Cath, T. Y. Osmotically and thermally driven membrane processes for enhancement of water recovery in desalination processes. *Desalin. Water Treat.* **2010**, *15*, 279–286.

32. Zhao, S., Zou, L., Tang, C. Y., Mulcahy, D. Recent developments in forward osmosis: Opportunities and challenges. *J. Membr. Sci.* **2012**, *396*, 1–21.

33. Greenlee, L. F., Testa, F., Lawler, D. F., Freeman, B. D., Moulin, P., Ce, P. The effect of antiscalant addition on calcium carbonate precipitation for a simplified synthetic brackish water reverse osmosis concentrate. *Water Res.* **2010**, *44*, 2957–2969.

34. Greenlee, L. F., Testa, F., Lawler, D. F., Freeman, B. D., Moulin, P. Effect of antiscalant degradation on salt precipitation and solid/liquid separation of RO concentrate. *J. Membr. Sci.* **2011**, *366*, 48–61.

35. McCool, B. C., Rahardianto, A., Cohen, Y. Antiscalant removal in accelerated desupersaturation of RO concentrate via chemically-enhanced seeded precipitation (CESP). *Water Res.* **2012**, *46*, 4261–4271.
36. Pérez-González, A., Urtiaga, M., Ibáñez, R., Ortiz, I. State of the art and review on the treatment technologies of water reverse osmosis concentrates. *Water Res.* **2012**, *46*, 267–283.
37. Halevy, S., Korin, E., Gilron, J., Box, P. O. Kinetics of gypsum precipitation for designing interstage crystallizers for concentrate in high recovery reverse osmosis. *Ind. Eng. Chem. Res.* **2013**, *52*, 14647–14657.
38. Subramani, A., Jacangelo, J. G. Treatment technologies for reverse osmosis concentrate volume minimization: A review. *Sep. Purif. Technol.* **2014**, *122*, 472–489.
39. Hann, F. French Patent No. 846,628, Nov 15, 1938.
40. Kaufman, C. E. Treatment of Water for Boiler Feed. U.S. Patent 2,395,331, Feb 19, 1946.
41. Klein, G., Villena-Blanco, M., Vermeulen, T. Ion-exchange equilibrium data in the design of a cyclic sea water softening process. *I&EC Process Des. Dev.* **1964**, *3* (3), 280–287.
42. Klein, G., Cherney, S., Ruddicks, E. L., Vermeulen, T. Calcium removal from sea water by fixed-bed ion exchange. *Desalination.* **1968**, *4* (2), 158–166.
43. United States Department of the Interior Office of Saline Water Research and Development. *Seawater Softening by Ion Exchange as a Saline Water Conversion Pretreatment.* Progress Report 62. 1962. http://digital.library.unt.edu/ark:/67531/metadc11622/m1/12/, accessed on October 10, 2014.
44. van Hoek, C., Kaakinen, J. W., Haugseth, L. A. Ion exchange pretreatment using desalting plant concentrate for regeneration. *Desalination.* **1976**, *19* (1–3), 471–479.
45. Kaakinen, J. W., Eisenhauer, R. J., van Hoek, C. High recovery in the Yuma desalting plant. *Desalination.* **1977**, *23* (1–3), 357–366.
46. Wilf, M., Konstantin, M., Chencinsky, A. Evaluation of an ion exchange system regenerated with seawater for the increase of product recovery of reverse osmosis brackish water plant. *Desalination.* **1980**, *34*, 189–197.
47. Vermeulen, T., Tleimat, B. W., Klein, G. Ion-exchange pretreatment for scale prevention in desalting systems. *Desalination.* **1983**, *47* (1–3), 149–159.
48. Barba, D., Brandani, V., Foscolo, P. U. A method based on equilibrium theory for a correct choice of a cationic resin in sea water softening. *Desalination.* **1983**, *48* (2), 133–146.
49. Shain, P., Klein, G., Vermeulen, T. A. Mathematical model of the cyclic operation of desalination-feedwater softening by ion-exchange with fluidized-bed regeneration. *Desalination.* **1988**, *69* (3), 135–146.
50. Muraviev, D., Khamizov, R., Tikhonov, N. A., Morales, J. G. Clean ("green") ion-exchange technologies. 4. High-Ca-selectivity ion-exchange material for self-sustaining decalcification of mineralized waters process. *Ind. Eng. Chem. Res.* **2004**, *43*, 1869–1874.
51. Tokmachev, M. G., Tikhonov, N. A., Khamizov, R. K. Investigation of cyclic self-sustaining ion exchange process for softening water solutions on the basis of mathematical modeling. *React. Funct. Polym.* **2008**, *68*, 1245–1252.
52. Mukhopadhyay, D. Method and Apparatus for High Efficiency Reverse Osmosis Operation. U.S. Patent 6,537,456, March 25, 2003.
53. Sarkar, S., SenGupta, A. K. A new hybrid ion exchange-nanofiltration (HIX-NF) separation process for energy-efficient desalination: Process concept and laboratory evaluation. *J. Membr. Sci.* **2008**, *324*, 76–84.
54. Venkatesan, A., Wankat, P. C. Desalination of the Colorado River water: A hybrid approach. *Desalination.* **2012**, *286*, 176–186.

55. Abdulgader, H. A., Kochkodan, V., Hilal, N. Hybrid ion exchange—Pressure driven membrane processes in water treatment: A review. *Sep. Purif. Technol.* **2013**, *116*, 253–264.
56. Helfferich, F. *Ion Exchange*, Dover, New York, 1995.
57. Clifford, D., Weber, W. J., Jr. The determinants of divalent/monovalent selectivity in anion exchangers. *React. Polym.* **1983**, *1*, 77–89.
58. Boari, G., Liberti, L., Merli, C., Passino, R. Exchange equilibria on anion resins. *Desalination.* **1974**, *15* (2), 145–166.
59. Aveni, A., Boari, G., Liberti, L., Santori, M., Monopoli, B. Sulfate removal and dealkalization on weak resins of the feed water for evaporation desalting plants. *Desalination.* **1975**, *16* (2), 135–149.
60. *OLI Stream Analyzer*, Version 9.0, OLI Systems, Cedar Knolls, NJ, **2013**.
61. Gregor, H. P. Gibbs-Donnan equilibrium in ion exchange resin systems. *J. Am. Chem. Soc.* **1951**, *73* (2), 642–650.
62. Gregor, H. P., Abolafia, O. R., Gottlieb, M. H. Ion-exchange resins. X. Magnesium-potassium exchange with a polystyrenesulfonic acid cation-exchange resin. *J. Phys. Chem.* **1954**, *58* (11), 984–986.
63. *Standard Methods for the Examination of Water and Wastewater*, 18th ed., American Public Health Association, Washington, DC, 1992.
64. *USEPA SulfaVer 4 Method 8051*; Hach Company. DOC316.53.01135.
65. Padungthon, S., Li, J., German, M., SenGupta, A. K. Hybrid anion exchanger with dispersed zirconium oxide nanoparticles: A durable and reusable fluoride-selective sorbent. *Environ. Eng. Sci.* **2014**, *31*, 360–372.
66. Clifford, D. A. Ion exchange and inorganic adsorption. In *Water Quality and Treatment: A Handbook of Community Water Supplies*, 5th ed., Mays, L. W. (Ed.), American Water Works Association, McGraw-Hill, New York, **1999**, pp. 9.1–9.91.
67. SenGupta, A. K., Greenleaf, J. E. Arsenic in subsurface water: Its chemistry and removal by engineered processes. In *Environmental Separation of Heavy Metals: Engineering Processes*, SenGupta, A. K. (Ed.), CRC Press, Boca Raton, FL, 2002, pp. 265–306.
68. Muraviev, D. N., Khamizov, R. Ion-exchange isothermal supersaturation: Concept, problems, and applications. In *Ion Exchange and Solvent Extraction*, Vol. 16, SenGupta, A. K., Marcus, Y. (Eds.), Marcel Dekker, New York, 2004, pp. 119–210.

2 Removal of Boron from Water by Ion Exchange and Hybrid Processes

Idil Yilmaz Ipek, Enver Guler,
Nalan Kabay, and Mithat Yuksel

CONTENTS

Boron is one of the vital nutrients for human beings, plants, and animals. Nevertheless, it may lead to detrimental effects when its concentration exceeds the permissible limits in irrigation water. There is no straightforward and simple technology for boron removal. Some conventional and advanced methods are chemical precipitation, liquid–liquid extraction, reverse osmosis, and electrodialysis. Among them, ion exchange, membrane processes, and a combination of both, the so-called sorption–membrane filtration hybrid process, have recently gained much attention. The use of boron-selective chelating ion exchange resins is of great interest as boron can be efficiently removed from water by ion exchange process. In hybrid systems, the main benefit is the higher efficiency with reduced amount of sorbents compared to conventional fixed-bed techniques. This chapter summarizes the results obtained by the authors for the removal of boron by ion exchange and hybrid processes along with the detailed literature survey on this topic. Ion exchange studies mainly cover equilibrium, kinetics, and column-mode tests. The hybrid process studies consist of the use of commercial sorbents combined with filtration by submerged membrane modules. The results showed that both ion exchange and hybrid approaches could be successful methods for the removal of boron from water.

2.1 INTRODUCTION

Elemental boron is rarely found in nature, as it forms a number of complex compounds, such as boric acid, borate, perborate, and others. Much of the boron that is released into the environment by human activity is associated with agriculture and industry.[1] Boron compounds have been used in many different applications including fertilizers, insecticides, corrosion inhibitors in anti-freeze formulations for motor vehicles and other cooling systems, buffers in pharmaceutical and dyestuff production, bleaching solutions in paper industries and in detergents, and also as moderators in nuclear reactors where anthropogenic water-soluble boron compounds are often discharged to the aqueous environment.[2] Besides, boron is widely distributed in surface and groundwaters, occurring naturally or from anthropogenic contamination, mainly in the form of boric acid or borate salts.[3] Water contamination by boron is one of the most widespread environmental problems, since even a few parts per million present in irrigation water can stunt plant growth. The problem of water deboronation is the challenge both for countries with a deficiency of potable water and the rest of world's communities.[3,4]

Boron is widely distributed throughout the lithosphere as a low concentrated element. Due to its high affinity to oxygen, it usually appears in the form of borates. It is also possible to find boron in the form of polyborates, complexes with transition metals, and as fluoroborate compounds. However, boric acid is the mostly dominant species in surface and brackish water when the pH is less than 9. When the pH raises over this value, borate anion $B(OH)_4^-$ can be found in aqueous solutions. The following equation shows the equilibrium[5]:

$$B(OH)_3 + H_2O \leftrightarrow B(OH)_4^- + H^+$$

Boron is usually present in water as boric acid, a weak acid that dissociates according to the following equilibria[6]:

$$H_3BO_3 \leftrightarrow H^+ + H_2BO_3^- \qquad pK_a = 9.14$$

$$H_2BO_3^- \leftrightarrow H^+ + HBO_3^{2-} \qquad pK_a = 12.74$$

$$HBO_3^{2-} \leftrightarrow H^+ + BO_3^{3-} \qquad pK_a = 13.8$$

Its concentration is usually expressed as "total boron" (tB), which includes all species and is expressed in terms of the molecular weight of the boron atom[6]:

$$tB \leftrightarrow [H_3BO_3] + [H_2BO_3^-] + [HBO_3^{2-}] + [BO_3^{3-}] \quad \text{as mg/L}$$

As seen in Figure 2.1, at a pH of 7–8, the predominant species in water is boric acid in molecular form. At these pH values, the percentage of the nondissociated species H_3BO_3 is between 99.3% (pH 7) and 93.2% (pH 8) of total boron. In the standard test condition at pH 8, 93.2% of molecular boron is present as neutral H_3BO_3 and 6.8% as $H_2BO_3^-$. At higher pH values, the percentage of $H_2BO_3^-$ increases while that of H_3BO_3 decreases.[6]

Between pH values of about 6 and 11, and at high concentration (>0.025 mol/L), highly water-soluble polyborate ions such as $B_3O_3(OH)_4^-$, $B_4O_5(OH)_4^-$, and $B_5O_6(OH)_4^-$ are formed.[5]

Boron is widely distributed in nature and is naturally found in minerals, plants, rocks, coal, and natural waters. The potential sources of boron contamination in water resources are either anthropogenic such as pollution from sewage effluents, boron-enriched fertilizers, and land-fill leachates or natural such as volcanic activity, geothermal steam, water–rock interaction, sea water encroachment, mixing with fossil brines, and hydrothermal fluids.[7,8] Other origins of contamination by boron can be found in the misuse of residual water for irrigation, or abuse in the use of pesticides or fertilizers, all of which are made up of appreciable quantities of boron.[9] Various activities including mining, irrigated agriculture, and the production of glass, cleaning products, and agrochemicals can produce boron-contaminated wastewater.[10] Another source of contamination is leaking residual waters originating from drainage systems. The boron content in residual waters is around 3 mg B/L and originates from the perborates used in the formulation of detergents.[9]

It is difficult to determine the exact amount of boron that enters the air, soil, or water from many of these sources. The natural borate content of groundwater and surface water is usually small. The borate content of surface water can be

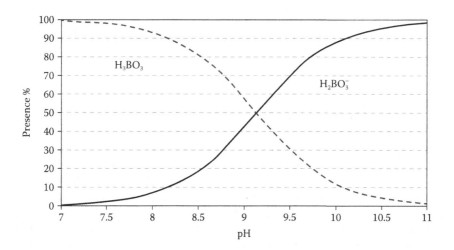

FIGURE 2.1 Distribution of $H_3BO_3/H_2BO_3^-$ as a function of pH. (Reprinted from *Desalination*, 156, Redondo, J., et al., Boron removal from seawater using FILMTEC™ high rejection SWRO membranes, 229–238, Copyright 2003, with permission from Elsevier.)

significantly increased as a result of wastewater discharges, because borate compounds are ingredients of domestic washing agents.[5] In surface water, concentrations of boron depend on the amount of boron present in the soils and rocks of the drainage area, and on the proximity of the drainage area to the ocean. The ocean provides boron both by depositing vaporized boric acid on the drainage area, and by infiltration of boron-containing seawater in tidal regions and estuaries. Surface waters can also receive boron inputs from effluent discharges, both from industrial processes and from municipal sewage treatment. Concentrations of boron in surface water range widely from 0.001 mg/L to as much as 360 mg/L. However, average boron concentrations are typically well below 0.6 mg/L.[11] Boron compounds are present in small amounts in seawater (5 mg/L) and in some mineral waters.[8] Naturally occurring boron is present in groundwater primarily as a result of leaching from rocks and soils containing borates and borosilicates. Concentrations of boron in groundwater throughout the world range widely from <0.3 to >100 mg/L.[5]

Boron is of special concern in irrigation water because of both its beneficial and toxic effects on plants.[12] Boron is an essential micronutrient related with the growth and development of plants. It is a structural component of cell walls that has been connected with the lignification processes, membrane transport, enzyme interactions, nucleic acid synthesis, and carbohydrate metabolism.[13] As a result, boron (in the form of borates) occurs naturally in fruits, nuts, and vegetables.[11] Its unique character is the exceptionally narrow concentration band in irrigation water, which is optimal for plant growth. A deficiency of boron will result in poor budding, excessive branching, and generally inhibited plant growth. A minimum of boron in irrigation water is required for certain metabolic activities, but if its concentration is only slightly higher, plant growth will exhibit effects of boron poisoning, which are yellowish spots on the leaves and fruits, accelerated decay, and ultimately plant expiration.[14] In fact, a level of boron in irrigation water exceeding 1 mg/L can significantly affect the yield of sensitive crops such as citrus fruits. For example, the optimal boron concentration range is 0.3–0.5 mg/L for citrus and grapes and 0.5–0.75 mg/L for corn. Continuous irrigation by desalinated seawater may increase boron concentrations in plant leaves by about three times the rate of increase caused by regular surface or underground water irrigation, due to high boron concentrations.[15]

The effect of boron has not been determined for animals as explicitly as was done for plants. It was shown that boron is an essential element in the human diet; however, its specific biochemical function has not been identified yet. Boron is important in the metabolism and utilization of calcium for human bone structuring. Other benefits of boron include improvement of brain function, psychomotor response, and the response to estrogen ingestion in postmenopausal women.[16] Boron in the elemental form is not toxic. The finely divided powder is hard and abrasive, and may cause skin problems indirectly if the skin is rubbed after contact. Humans who experience acute boron poisoning may be subject to the following symptoms: nausea, vomiting, diarrhea, headache, skin rashes, desquamation (the shedding of the outer layers of the skin), and central nervous system stimulation (leading to depression). Boron accumulated in the body through absorption, ingestion, and inhalation of its compounds affects the central nervous system.[17,18] High boron levels in drinking water can be toxic to humans as boron has been shown to cause male reproductive impediments in laboratory animals.[19,20]

A drinking water quality guideline value of 2.4 mg/L has been recently suggested by the World Health Organization (WHO). This limit value is well below boron contents found in water resources of many areas. Noteworthy, this value has been decided exclusively by extrapolating possible effects on human health. On the contrary, a reasonable arrangement among health remarks and performance of available water treatment technology was assumed as the criterion in the past. Curiously, the previous guideline value was 0.5 mg/L, far lower than the current one.[21]

2.2 OVERVIEW OF BORON REMOVAL METHODS

Potable and irrigation water demand increases whereas suitable water sources diminish inevitably. Thus, boron removal at low concentrations from natural sources and industrial effluents deserves serious attention.[4] Water is essential to the survival of humans, animals, and plants and has played a major role in human behavior throughout the history of mankind. Scarcity of water in particular regions led to forced migrations. The increasing demand for potable water is a pressing concern, and although Earth is composed mainly of water, only 1% of this is consumable fresh water.[22]

There are several methods for removing boron from water that contains high concentration of boron, even though most of them are inefficient to obtain the desired concentration and difficult to utilize in terms of disadvantages.[12] The methods for boron removal can be classified such as co-precipitation, evaporation–crystallization, solvent extraction, adsorption, hydrothermal treatment, processes based on liquid–liquid electrochemistry, electrocoagulation (EC), membrane filtration, ion exchange, and hybrid processes.

A co-precipitation method using metal hydroxide is an inefficient and environmentally ineffective process due to low removal rate, requirement of a large amount of metal hydroxide, and disposal of large amount of unrecyclable wastes.[23] The removal efficiency by co-precipitation is about 90% in the boron range of 0.16–1.6 mg B/L using $Al_2(SO_4)_3$ and $Ca(OH)_2$. This method is not effective due to the sludge production at the end of the process.[17] Lime is frequently utilized in coagulation process, but it can remove only 87% of boron from concentrated wastewater at 60°C.[24] The residual boron in the product water treated by chemical oxoprecipitation, which combines treatment with an oxidant and precipitation using metal salts at pH 10, was reached as 15 mg/L, which cannot be achieved using conventional coagulation processes.[1] Conventional sedimentation and biological treatment remove little boron from wastewater, and chemicals commonly used in the water treatment industry also have little or no effect on the boron levels in water.[12] Yilmaz et al. compared EC and chemical coagulation approaches for boron removal. At optimum conditions, boron removal efficiencies for EC and chemical coagulation were found to be 94% and 24%, respectively.[25] Elsewhere, boron removal from boron-containing wastewaters prepared synthetically via an EC method was studied by Yilmaz et al. They found that the efficiency of boron removal increased with increasing current density and decreased with increasing boron concentration in the solution.[26] In a similar study, Yilmaz et al. selected EC as a treatment process for the removal of boron from thermal waters. At optimum conditions, boron removal efficiency in geothermal water reached up to 95%.[27] A hydrothermal treatment technique was developed by Itakura et al.[23] to recover boron as a recyclable precipitate $Ca_2B_2O_5 \cdot H_2O$ from aqueous solutions. As a result, it was found that the hydrothermal treatment using calcium hydroxide as a mineralizer converted boron in the aqueous media effectively into calcium borate, $Ca_2B_2O_5 \cdot H_2O$. In the optimal hydrothermal condition, more than 99% of boron was collected from the synthetic wastewater of 500 mg B/L.[23] Pieruz et al. investigated a process based on liquid–liquid electrochemistry to transfer selectively borate ions present in produced water to an immiscible phase.[28] An evaporation–crystallization process is sufficient only in the case of streams with very high boron concentrations, especially those that contain more than several thousand milligrams of boron per liter.[23] Various aliphatic 1,3-diols, containing primary, secondary, tertiary, and a mix of these groups, were used for the solvent extraction of boron by Karakaplan et al.[29] They found the best extracting reagent was diol 11, which is a primary–tertiary class of −OH groups and methyl groups as a substituent on the second carbon of 1,3-diol.[29]

Adsorption of boron on different kinds of adsorbents such as activated sludge[30]; Siral 5, 40, and 80[17]; hydrous cerium oxide; hydrous zirconium oxide; coal; and fly ash[19] has been investigated by numerous researchers. Xu et al. synthesized a novel adsorbent as silica-supported N-methyl-D-glucamine (NMDG) adsorbent (Si-MG) for boron removal from an aqueous solution.[31] Demey et al. prepared a new adsorbent [chiFer(III)] using chitosan and iron(III) hydroxide for boron recovery from seawater; it was a new composite containing chitosan as the encapsulating material and nickel(II) hydroxide [chiNi(II)] for the removal of boron from aqueous solutions.[22,32] Zelmanov and Semiat synthesized Fe(III) oxide/hydroxide nanoparticles sol (NanoFe) and NanoFe-impregnated granular activated carbon (GAC) as adsorbents for boron removal from solutions. They obtained at least 95%–98% of boron removal efficiency using NanoFe sol and Fe-impregnated GAC.[2]

There are also membrane processes such as reverse osmosis (RO) and electrodialysis (ED) for boron removal from water. However, these techniques require high cost for the manufacturing and maintenance. In regions where the shortage of water is a concern, treatment processes have been developed to obtain water from seawater and brackish waters by means of RO processes. In most cases these processes present rejection of salts as high as 98%. However, due to the chemistry of boron, at a normal pH of operation of these plants, the boron is in the form of boric acid and its elimination is reported to be around 43%.[33] For RO, the removal efficiency was about 40%–80% at natural seawater pH and at pH 10–11 over 90%. This technique is not effective due to membrane stability, costs, and membrane scaling because of $CaCO_3$ precipitation. The reason for the low rejection of the boric acid is due to its ability to diffuse through the membranes in a nonionic way, similar to that of carbonic acid or water.[15] Moreover, owing to the predominance of the noncharged boric acid in the solution, only some of the boron (40%–65%) is removed by RO. However, a second-stage RO desalination under high pH condition increases boron rejection to 92%.[19]

Seawater contains around 4–6 mg B/L, and the boron concentration of the product water from conventional RO seawater desalination operating at 45%–50% recovery is around 1.6–2.0 mg/L, depending on the content in the feed, water temperature, and pH. Recently developed spiral-wound membranes with improved boron rejection operating in two stages (with 55% of recovery) can produce permeate with 0.79–0.86 mg/L boron concentration.[34] Redondo et al. tested boron removal from seawater using FILMTEC™ high rejection spiral-wound RO (SWRO) membranes. According to the results, a shift to pH 10 brings the total boron rejection from 93% to 99% depending on the membrane chemistry. At a pH of 11, the total boron rejection is 99.0%–99.5%. Based on experimental observations, it is obvious that high pH operation is advantageous for boron rejection.[6] Ozturk et al. obtained 69% rejection of boron from a solution of 5 mg B/L at pH 9 using RO.[35] Conventional ED is only capable of removing about 42%–75% of boron. The effectiveness of boron removal by ED was shown to be dependent on the pH of the solution.[36] Kabay et al. applied an ED method for boron removal from aqueous solutions to investigate the effect of feed characteristics and interfering ions such as chloride and sulfate.[37] Yavuz et al. investigated boron removal from geothermal water using a mini pilot-scale RO system containing two spiral-wound FILMTEC™ BW30-2540 membranes, two types of FILMTEC™ SWRO membranes (SW30-2540 and a novel high rejection SWRO membrane abbreviated as XUS SW30XHR-2540) installed at a geothermal area.[8,38] They also studied the effect of pH on boron removal from geothermal water by using a mini pilot-scale brackish water RO system. The maximum boron rejection was found to be 47% at 12 bar of applied pressure while it increased to 49% at 15 bar of applied pressure using a single membrane mode of operation. When the pH of geothermal water was increased to 10.5, 94.5%–95% boron rejection was obtained for both single- and double-membrane configurations at 12 bar of operating pressure.[39] Guler et al. presented the field performance of a small-scale seawater RO unit employing polyamide thin-film composite spiral-wound seawater RO membranes (FILMTEC™ XUS SW30XHR-2540 and FILMTEC™ SW30-2540).[40] A hybrid process coupling RO with electrodeionization (EDI) was applied to remove boron and silica from geothermal water by Arar et al. The obtained results showed that it was not possible to lower the concentration of boron in the product water to the permissible level for irrigation water by means of an RO process at the natural pH of geothermal water. When the RO permeate collected was passed through an EDI cell, it was possible to decrease the concentration of boron to below 0.5 mg B/L at the end of the EDI operation when the optimal conditions were employed.[41] Oner et al. applied a laboratory-scale cross-flow flat sheet RO method for boron, salt, and silica removal from geothermal water by using four different RO membranes: AD-SWRO (GE Osmonics), AG-BWRO (GE Osmonics), BW-30-BWRO (FILMTEC™), and AK-BWRO (GE Osmonics).[42]

In the Middle East and North Africa, RO has been used to provide a large amount of drinking water and irrigation water.[43] However, boron removal by conventional RO technology is not always sufficient to meet these requirements for drinking water consumption and irrigation. Although the rejection rate of the conventional RO membranes for both anions and cations was almost 100%, the rejection of boron was only about 70%, resulting in 1–2 mg/L of boron in the product water (Figure 2.2).[44,45] This level of boron is not acceptable when standards for

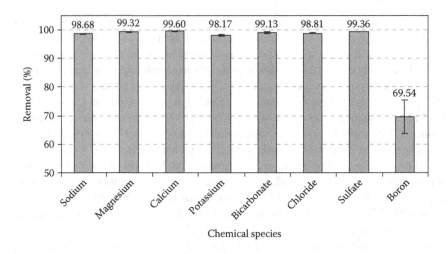

FIGURE 2.2 Removal percentages of chemical species by an RO system. (With kind permission from Springer Science+Business Media: *Environmental Geochemistry and Health*, Effect of temperature on seawater desalination-water quality analyses for desalinated seawater for its use as drinking and irrigation water, *32*, 2010, 335–339, Guler, E., et al.)

boron in irrigation water are considered. Therefore, over the last decades, novel water treatment processes, in terms of boron removal, have taken significant attention due to trace amount of boron in water streams (e.g., seawater and geothermal water) and strict water quality regulations. In that sense, interest in hybrid processes combining both ion exchange (i.e., sorption) and membrane filtration increased significantly.

To overcome the problem of having high concentrations of boron in RO permeate, polishing the water stream has been recently considered as a post-treatment using boron-specific ion exchange resins (or sometimes referred as boron-selective resins, or BSR).[12,43,45–48] A chelating type of resin has been used to remove boron efficiently. It shows a high selectivity to boron while these groups are not reactive to ordinary metals and other elements. Commercially available BSR contain NMDG groups that capture boron and make complexes via covalent attachment.[48] Formation of boron complexes by sorbents and subsequent filtration by membrane processes has been suggested as an alternative to remove and/or recover boron under relatively low operating pressures.[49–51] Such hybrid processes specifically for boron removal purposes have not yet been implemented on an industrial scale although up-to-date results are quite promising. Ion exchange processes using resins have some advantages such as the operation is simple and no energy is required for exchanging ions of one to another. However, regeneration and the ion exchanging capacity of those materials should always be taken into account before implementation in practice. For membrane filtration, low-energy membrane filtration processes can be considered such as microfiltration (MF) and ultrafiltration (UF).

2.3 REMOVAL OF BORON FROM WATER BY ION EXCHANGE

Among several methods of boron removal, the use of boron-selective chelating ion exchange resins still seems to have the greatest importance. Ion exchange processes are efficiently used to remove undesirable ions, such as arsenic, boron, and heavy metal ions, from water.

The predominate boron species (above 90%) in water (at neutral pH 7.7) is $B(OH)_3$. Boric acid reacts readily with polyalcohols to form an acid considerably stronger than the original boric acid. Although boric acid cannot be conveniently titrated with sodium hydroxide, the addition of an excess of mannitol permits one to titrate boric acid with a base using phenolphthalein. Kunin and Preuss have used this principle in synthesizing a boron-selective ion exchange resin by aminating a chloromethylated styrene–divinylbenzene copolymer with NMDG.[52] Commercially available boron-selective ion exchange resins are based on the macroporous polystyrene matrix. After chloromethylation and amination with NMDG, boron-selective ion exchange resins are produced. The structure of boron-selective chelating ion exchange resins is shown in Figure 2.3.

The presence of two vicinal hydroxylic groups allows boric acid and borates to form stable complexes. The idea of complexation is shown in Figure 2.4.[53,54] The process is based on a weak base (tertiary amine) that interacts with the uptake of a boron as borate $[B(OH)_4]^-$ leading to the formation of a complex. The extreme selectivity of the resin makes this ion exchange process highly efficient. Unlike with other processes such as RO, only the targeted contaminant, boron, is removed selectively from the water while sodium, calcium, chloride, and bicarbonate ions are not eliminated.

Many compounds with vicinal diol groups, such as tiron, chromotropic acid, and NMDG, are usually considered to be the efficient ligands for the boron chelation.[16,31] However, NMDG is always regarded as the most efficient ligand. Polymer-supported NMDG chelating resins, such as Amberlite IRA 743, Diaion CRB 02 and D564,[31] Purolite S108, Diaion CRB 01, Diaion CRB 02,[55,56] Diaion CRB 02, Dowex (XUS 43594.00) and Purolite S 108,[57] Diaion CRB 03, Diaion CRB 05,[58]

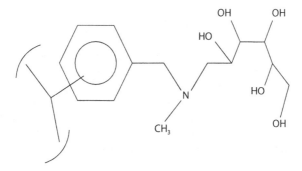

FIGURE 2.3 The structure of boron-selective chelating resin containing NMDG group. (From Busch, M., et al., A boron selective resin for seawater desalination, *Proceedings of European Desalination Society Conference on Desalination and Environment*, Santa Margherita, Italy, 2005. With permission.)

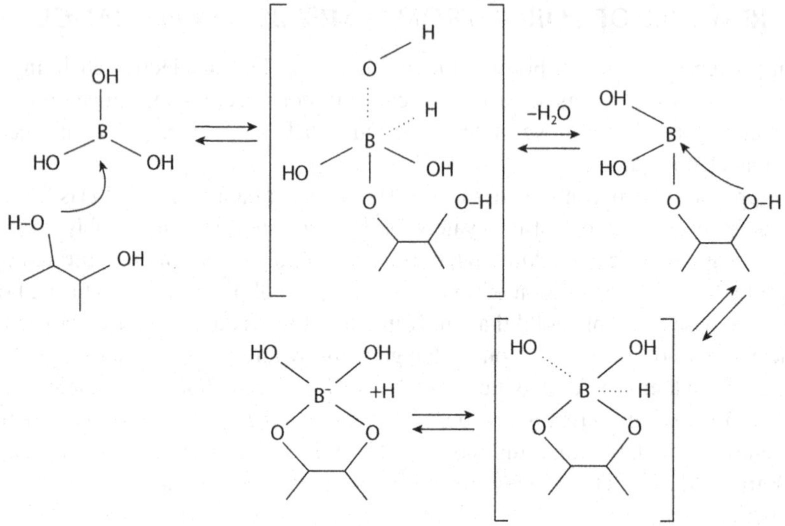

FIGURE 2.4 Boric acid complexation with vicinal –OH groups. (From Busch, M., et al. A boron selective resin for seawater desalination, *Proceedings of European Desalination Society Conference on Desalination and Environment*, Santa Margherita, Italy, 2005. With permission.)

a novel boron sorbent produced by grafting a boric acid chelating group, NMDG, onto the hydrophilic silica–polyallylamine composites,[59] a sorbent fabricated by the functionalization of a natural biopolymer, chitosan, with NMDG through atom transfer radical polymerization,[60] poly(propylene) sheet sorbent chemically modified with NMDG and functionalized in situ crosslinking of poly(vinylbenzyl chloride) with a cyclic diamine piperazine,[61] have been made for separation of boron from water.

Additionally, there are some other synthetic and natural ion exchange materials such as anion exchange resins prepared by the reaction of chloromethylated poly(styrene-*co*-divinylbenzene) with 1-amino-1-deoxy-D-glucitol, 1-deoxy-1-(methylamino)-D-glucitol, 1-1-iminobis(1-deoxy-D-gluciol), and 1,4-piperazinediylbis(1-deoxy-D-glucitol),[62] macroporous chelating resins containing polyol groups (RGB),[63] anion exchange resins impregnated with citric acid and tartaric acids,[64] synthesized *N*-methylglucamine–type cellulose derivatives,[65] and Lewatit MK 51.[66]

Recently, there has been great interest in synthesizing novel BSR for removing boron from geothermal water, seawater, and RO permeate. Santander et al. synthesized a novel chelating resin poly(*N*-(4-vinylbenzyl)- *N*-methyl-D-glucamine), P(VbNMDG), and compared its sorption performance with boron-selective commercial resin Diaion CRB02 containing NMDG groups for boron removal from geothermal water. The P(VbNMDG) resin showed a higher sorption capacity and faster kinetics than that of Diaion CRB02 for boron removal from geothermal water.[67] Wolska et al. prepared boron-selective particles reacting NMDG with poly(styrene–vinylbenzyl chloride–divinylbenzene).[68,69] Samatya et al. obtained promising results using monodisperse porous particles with dextran-based molecular brushes attached to the particles via "click chemistry" and "direct coupling,"[70] monodisperse porous poly(glycidyl methacrylate-*co*-ethylene methacrylate) particles carrying diol functionality,[71,72] and monodisperse porous poly(vinylbenzyl chloride-*co*-divinylbenzene) beads 8.5 μm in size that are synthesized by a new "modified seeded polymerization" technique[73] for boron removal. Zerze et al. synthesized a copolymer, poly(vinyl amino-*N*,*N'*-*bis*-propane diol-*co*-DADMAC) (GPVA-*co*-DADMAC), to remove boron via polymer-enhanced UF (PEUF) system.[4] Simsek et al. used two selective resins, an NMDG-modified poly(styrene)-based polymer (VBC–NMG) and an iminodipropylene glycol functionalized polymer (GMA–PVC), for boron removal. These novel BSR depicted promising results for practical uses as they could be used for more than eight cycles of recycling without any significant change in sorption capacity.[74] Wang et al. developed a new kind of chelating resin with a pyrocatechol functional group for boron removal from aqueous solutions in broad pH conditions.[75]

Besides these investigations on boron removal, boron-selective ion exchange resins Diaion CRB 02 and Dowex (XUS 43594.00) have been employed to examine boron removal from geothermal water.[76] The optimal amount of resin was determined to be 4.0 g resin/L geothermal water for geothermal water containing around 20 mg B/L when both Diaion CRB 02 and Dowex (XUS) resins at a particle size range of 0.355–0.500 mm were employed (Figure 2.5). The kinetic behaviors of Diaion CRB 02 and Dowex (XUS) at a particle size of 0.355–0.500 mm were compared in order to get a measure of the relative kinetic performance of the resins at different stirring rates of 200, 250, and 300 rpm. Figures 2.6 and 2.7 show that the stirring rate has no effect on the kinetic performance of both resins for removing boron from geothermal water.

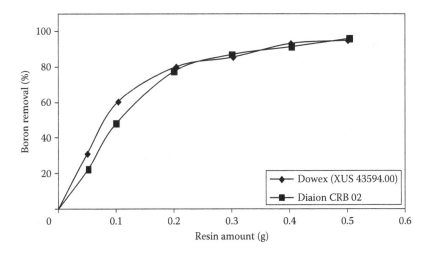

FIGURE 2.5 Effect of resin amount on boron removal from geothermal water (resin amount is g/100 mL). (Adapted from Yilmaz-Ipek, I., *Boron removal from geothermal water by ion exchange-membrane hybrid process*, PhD Thesis, Ege University, 2009. With permission.)

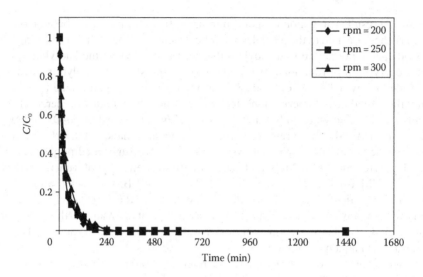

FIGURE 2.6 Effect of stirring rate on boron removal from geothermal water by Diaion CRB 02. (Adapted from Yilmaz-Ipek, I., *Boron removal from geothermal water by ion exchange-membrane hybrid process*, PhD Thesis, Ege University, 2009. With permission.)

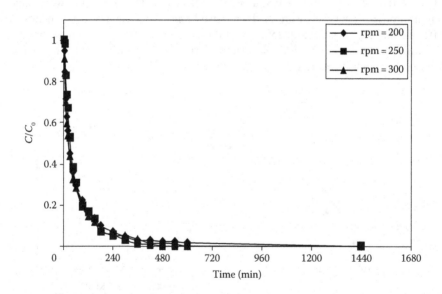

FIGURE 2.7 Effect of stirring rate on boron removal from geothermal water using Dowex (XUS 43594.00). (Adapted from Yilmaz-Ipek, I., *Boron removal from geothermal water by ion exchange-membrane hybrid process*, PhD Thesis, Ege University, 2009. With permission.)

To understand the mechanism of sorption for boron removal from geothermal water using boron-selective ion exchange resins and the sorption rate controlling step, all the kinetic data were evaluated using pseudo-first-order and pseudo-second-order kinetics models. A simple kinetic model of sorption is the pseudo-first-order equation in the form of[66]

$$\frac{dq_t}{dt} = k_1 (q_e - q_t) \tag{2.1}$$

Integrating Equation 2.1 and applying the boundary conditions $q_t = 0$ at $t = 0$ and $q_t = q_t$ at $t = t$, gives

$$\log(q_e - q_t) = \log q_e - \frac{k_1}{2.303} t \tag{2.2}$$

where q_e and q_t are the amounts of boron sorbed at equilibrium and at time t (mg/g), respectively, and k_1 is the rate constant of pseudo-first-order sorption (min^{-1}).

A pseudo-second-order equation based on sorption equilibrium capacity may be expressed in the form[66]

$$\frac{dq_t}{dt} = k_2 (q_e - q_t)^2 \tag{2.3}$$

After defining integration by applying the initial conditions, Equation 2.3 becomes

$$\frac{1}{(q_e - q_t)} = \frac{1}{q_e} + k_2 t \tag{2.4}$$

Equation 2.4 can be rearranged to obtain a linear form:

$$\frac{t}{q_t} = \frac{1}{k_2 q_e^2} + \frac{t}{q_e} \tag{2.5}$$

where k_2 is the rate constant of the pseudo-second-order sorption (g/mg min).

To obtain the correlation coefficients, $\log(q_e - q_t)$ versus t and t/q_t versus t were plotted for first-order and second-order kinetic models, respectively. Since the correlation coefficients of pseudo-second-order kinetics (R^2) are mostly greater than those of pseudo-first-order kinetics for almost all stirring rates, it can be concluded that the sorption of boron by both resins obeys the pseudo-second-order kinetic model.

To examine the rate-controlling mechanism of the sorption process, several kinetic models are used to test the experimental data. The models for process dynamics include both the diffusional steps (bulk solution, a film layer at the external surface of the particle, pores) and the exchange reaction on the active sites. Since the resistance in bulk solution is easily controlled and negligible, three resistances, such as film diffusion, particle diffusion, and chemical reaction, usually determine the overall rate of the ion exchange process. One approach uses the infinite solution volume (ISV) model, whereas the other method uses the unreacted core model (UCM) to describe the rate-determining steps in the ion exchange process.[66] These two kinetic models developed for spherical particles to specify the rate-determining steps are given in Table 2.1.

The maximum correlation coefficients for the linear models show that the boron sorption by both resins is controlled by particle diffusion according to the ISV and UCM models for all stirring rates.

The effects of resin particle size on sorption kinetics behaviors of Diaion CRB 02 and Dowex (XUS) boron-selective ion exchange resins for removing boron from geothermal water are compared in Figure 2.8. Sorption kinetics of both resins at a smaller particle size range (0–0.045 mm) are much more rapid than that of the same resin

---

TABLE 2.1
Diffusional and Reaction Models

Model	Equation	Rate-Determining Step
ISV	$F(X) = -\ln(1 - X) = K_{1i}t$, where $K_{1i} = 3DC/r_o\delta C_r$	Film diffusion
ISV	$F(X) = -\ln(1 - X^2) = kt$, where $k = D_r\pi^2/r_o^2$	Particle diffusion
UCM	$F(X) = X = (3C_{Ao}K_{mA}/a_{ro}C_{so})t$	Liquid film
UCM	$F(X) = 3 - 3(1 - X)^{2/3} - 2X = (6D_{eR}C_{Ao}/a_{ro}^2C_{so})t$	Reacted layer
UCM	$F(X) = 1 - (1 - X)^{1/3} = (k_sC_{Ao}/a_{ro}C_{so})t$	Chemical reaction

Source: Adapted from Yilmaz-Ipek, I., *Boron removal from geothermal water by ion exchange-membrane hybrid process*, PhD Thesis, Ege University, 2009. With permission.

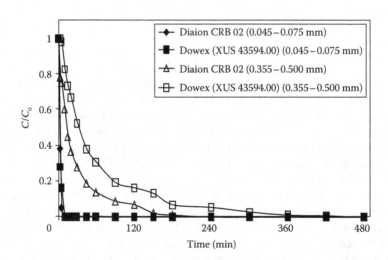

FIGURE 2.8 Effect of resin particle size on sorption kinetics performance of boron-selective ion exchange resins. (Adapted from Yilmaz-Ipek, I., *Boron removal from geothermal water by ion exchange-membrane hybrid process*, PhD Thesis, Ege University, 2009. With permission.)

at a particle size range of 0.355–0.500 mm since the ability of boron to reach the binding sites of the resin increased due to increased effective surface area.

The effect of pH on boron removal from geothermal water was investigated using Dowex (XUS) boron-selective ion exchange resin. Boron removal percentages were found to be higher at pHs 9 and 10, which are 91% and 93%, respectively, higher than those at pH 8.4, which is 85%. It was considered that the complex formation between boron in the solution and *cis*-diol groups on the resin should occur at a certain pH. The effective pH range for boron sorption by Dowex (XUS) BSR is given between 6 and 10 by the resin manufacturer. Indeed, boron removal percentage decreased at pH 11 (Figure 2.9).

Additionally, the kinetic behavior of Diaion CRB 02 ion exchange resin for boron removal from geothermal water was examined to get a measure of the relative kinetic performance of the resins as a function of temperature. Figure 2.10 shows that at a moderate stirring rate, the boron removal rate was found to be higher at higher temperatures. It can be concluded that the diffusion of boron to the functional groups of the resin was improved due to the increase in temperature.

The kinetic data obtained at different temperatures were evaluated using pseudo-first-order and pseudo-second-order kinetic models. The linearity of the plots shows that sorption kinetics obey the second-order kinetic model at 25, 35, and 45°C.

The maximum correlation coefficients for the diffusional and reaction models show that the rate-determining step is particle diffusion controlled according to both the ISV and UCM models at 25°C, 35°C, and 45°C for Diaion CRB 02 resin.

Thermodynamic parameters such as standard free energy change ($\Delta G°$), standard enthalpy change ($\Delta H°$), and standard entropy change ($\Delta S°$) should be determined at various temperatures to evaluate the feasibility and the effect of temperature better for boron sorption using boron-selective ion exchange resins. The Gibbs free energy

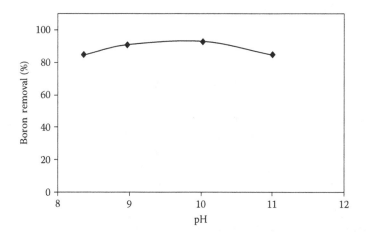

FIGURE 2.9   Effect of pH on boron removal from geothermal water. (Adapted from Yilmaz-Ipek, I., *Boron removal from geothermal water by ion exchange-membrane hybrid process*, PhD Thesis, Ege University, 2009. With permission.)

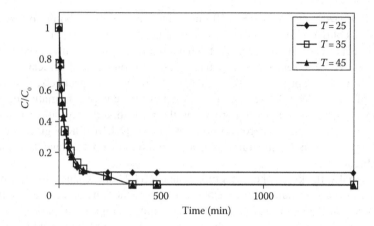

FIGURE 2.10 Effect of temperature on boron removal from geothermal water by Diaion CRB 02. (Adapted from Yilmaz-Ipek, I., *Boron removal from geothermal water by ion exchange-membrane hybrid process*, PhD Thesis, Ege University, 2009. With permission.)

change of sorption process is related to the equilibrium constant (K_c) and calculated using the following equations[18,77]:

$$\Delta G^\circ = -RT \ln K_c \tag{2.6}$$

$$K_c = C_s/C_e \tag{2.7}$$

where C_s is the amount of boron adsorbed (mmol/g), C_e is the equilibrium concentration (mmol/L) of boron in the solution, T is the solution temperature (K), and R is the gas constant (8.314 J/mol K).

The standard enthalpy change ΔH° and ΔS° values of sorption can be calculated from the Van't Hoff equation given as[18,77]:

$$\ln K_c = \frac{\Delta S^\circ}{R} - \frac{\Delta H^\circ}{RT} \tag{2.8}$$

The values of ΔH° and ΔS° are estimated from the slope and intercept of the plot of $\ln K_c$ against $1/T$. ΔG° values were calculated as −4.456, −5.626, and −5.904 kJ/mol at 298, 308, and 318 K, respectively. The negative values of ΔG° at given temperatures indicate the spontaneous nature of the sorption (Table 2.2). The negative values of ΔG° confirm the feasibility of the sorption process. It is known that the absolute magnitude of the change in free energy for physisorption is between −20 and 0 kJ/mol; chemisorption has a range of −80 to −400 kJ/mol.[77] In our case, the results show that the sorption of boron using both resins occurs as physically. The positive value of standard enthalpy change for boron sorption implies the endothermic nature of the sorption process. Positive values of ΔS° indicate an increase in degree of freedom at the solid/solution interface during boron sorption using Diaion CRB 02.

TABLE 2.2

Thermodynamic Parameters for Sorption of Boron Using Diaion CRB 02

Temperature (K)	K_c (L/g)	$\Delta G°$ (kJ/mol)	$\Delta H°$ (kJ/mol)	$\Delta S°$ (kJ/mol K)
298	6.04	−4.456		
308	9.00	−5.626	16.325	0.070
318	9.33	−5.904		

Source: Adapted from Yilmaz-Ipek, I., *Boron removal from geothermal water by ion exchange-membrane hybrid process*, PhD Thesis, Ege University, 2009. With permission.

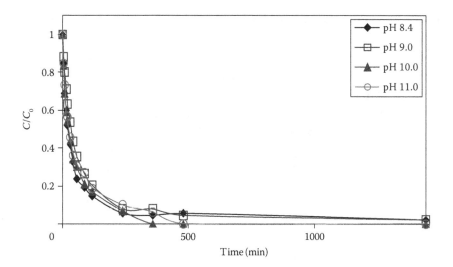

FIGURE 2.11   Effect of pH on boron removal from geothermal water by Diaion CRB 02. (Adapted from Yilmaz-Ipek, I., *Boron removal from geothermal water by ion exchange-membrane hybrid process*, PhD Thesis, Ege University, 2009. With permission.)

The kinetic behaviors of Diaion CRB 02 and Dowex (XUS) ion exchange resins for boron removal from geothermal water were also examined to get a measure of the relative kinetic performance of the resins as a function of pH. Figure 2.11 shows that even though it is not easy to determine the optimum pH value for Diaion CRB 02, the boron removal rate is more favorable at pHs 8.4 and 10 than at pHs 9 and 11 due to the ratio of boron concentration to the initial concentration, which was obtained as 0.50 at 20, 30, 20, and 30 min of contact time at pHs 8.4, 9.0, 10.0, and 11.0, respectively. The sorption rate obtained using Dowex (XUS) was found to be higher at pHs 9 and 10 than at pHs 8.4 and 11 as shown in Figure 2.12.

According to conventional kinetic models, it can be concluded that the sorption kinetic data obeyed second-order kinetic model using both Diaion CRB 02 and

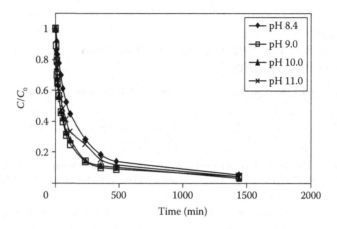

FIGURE 2.12 Effect of pH on boron removal from geothermal water by Dowex (XUS 43594.00). (Adapted from Yilmaz-Ipek, I., *Boron removal from geothermal water by ion exchange-membrane hybrid process*, PhD Thesis, Ege University, 2009. With permission.)

TABLE 2.3
Effect of pH on Column-Mode Boron Removal from Geothermal Water

pH	Breakthrough Capacity (mg B/mL resin)	Total Capacity (mg B/mL resin)	Elution Efficiency (%)
8.4	2.02	4.60	88.5
9.0	2.53	4.80	81.9
10.0	2.12	4.66	82.7
11.0	1.72	4.28	95.7

Source: Adapted from Yilmaz-Ipek, I., *Boron removal from geothermal water by ion exchange-membrane hybrid process*, PhD Thesis, Ege University, 2009. With permission.

Dowex (XUS) at each pH, whereas it is not easy to evaluate the data obtained at pH 11.0 for Dowex (XUS) since the correlation coefficients for the two models are very close to each other.

The maximum correlation coefficients for the diffusional and reaction models show that the rate is particle diffusion controlled according to both the ISV and UCM models at pHs 8.4, 10.0, and 11.0 using Diaion CRB 02, whereas such result was obtained at pHs 8.4, 9.0, and 10.0 using Dowex (XUS) (Table 2.3). However, the rate-determining step at pH 9 for Diaion CRB 02 and at pH 11 for Dowex (XUS) was film diffusion according to ISV and chemical reaction according to the UCM model.

In order to reuse an ion exchange resin several times, it is very crucial that the sorbed species be eluted from the resin quantitatively and the resin can be recycled many times after a regeneration process. In order to obtain a high elution efficiency, acid eluants such as H_2SO_4 and HCl at various concentrations were used for stripping boron from Diaion CRB 02 boron-selective ion exchange resin after sorption of boron from geothermal water. As shown in Figures 2.13 and 2.14, a quantitative stripping of boron was obtained with a percentage of 95%–100% even with low concentrations of H_2SO_4 and HCl such as 0.025 and 0.05 M, respectively.

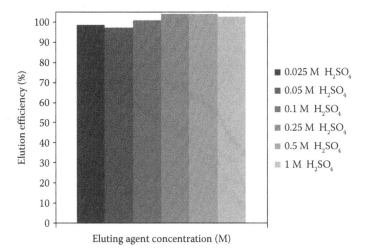

FIGURE 2.13 Effect of eluting agent concentration for stripping of boron from Diaion CRB 02 boron-selective ion exchange resin by H_2SO_4. (Adapted from Yilmaz-Ipek, I., *Boron removal from geothermal water by ion exchange-membrane hybrid process*, PhD Thesis, Ege University, 2009. With permission.)

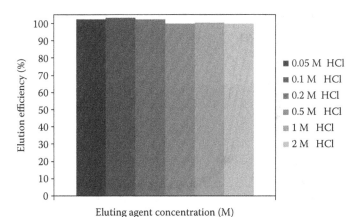

FIGURE 2.14 Effect of eluting agent concentration for stripping of boron from Diaion CRB 02 boron-selective ion exchange resin by HCl. (Adapted from Yilmaz-Ipek, I., *Boron removal from geothermal water by ion exchange-membrane hybrid process*, PhD Thesis, Ege University, 2009. With permission.)

In order to get some information about column performances of chelating ion exchange resin on removal of boron from geothermal water at different pHs, a chelating resin Dowex (XUS 43594.00) was used for column-mode sorption of boron. Breakthrough curves of boron obtained at different pHs such as 8.4, 9.0, 10.0, and 11.0 are given in Figure 2.15. The breakthrough capacity of Dowex (XUS 43594.00) resin was found to be the highest at 2.53 mg B/mL resin at pH 9.0 (Table 2.3).

Elution profiles of boron using Dowex (XUS 43594.00) resin are given in Figure 2.16. As tabulated in Table 2.3, boron loaded resins were eluted efficiently with 5% H_2SO_4 at SV 10 h^{-1}.

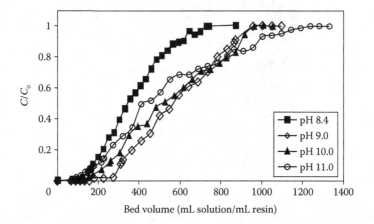

FIGURE 2.15 Effect of pH on breakthrough curves (SV = 20 h^{-1}). (Adapted from Yilmaz-Ipek, I., *Boron removal from geothermal water by ion exchange-membrane hybrid process*, PhD Thesis, Ege University, 2009. With permission.)

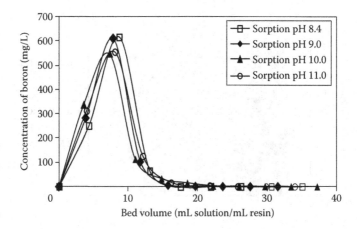

FIGURE 2.16 Elution profiles of boron (SV = 10 h^{-1}). (Adapted from Yilmaz-Ipek, I., *Boron removal from geothermal water by ion exchange-membrane hybrid process*, PhD Thesis, Ege University, 2009. With permission.)

2.4 REMOVAL OF BORON FROM WATER BY HYBRID PROCESSES

2.4.1 HYBRID PROCESSES USING WATER-SOLUBLE COMPLEXING AGENTS

Hybrid systems may use either soluble or insoluble sorbents. Geffen et al.[78] investigated soluble sorbents. By adding mannitol, di-borate esters have been formed which provided 90% boron removal at pH 9.0 by NF. Feed concentrations of both 7 and 32 mg/L have been investigated, and the impact of pH on the performance of boron removal has been detailed.

Dydo and Turek,[79] however, concentrated more on the recovery of boron using D-mannitol in an ED–RO system such that boron and polyol (D-mannitol) containing water is first desalinated by ED, and then after alkalization by RO in a two-stage configuration. The first RO stage produced 1 mg/L of boron in the permeate, and in the second stage a boron-rich alkaline solution is concentrated to a retentate boron content of around 10 mg/L or more. In the end, this stream has been neutralized and crystallized to obtain solid boric acid. In another study of Dydo et al.[80] borate complexes were removed by a direct RO method in the presence of polyols. The study presented the best membranes that could achieve this separation and the optimum operating conditions.

Using the advantage of complexation by polyols, but this time with nanofiltration (NF) and RO membranes, boron rejection was investigated.[81] Glycerol, mannitol, and sorbitol were used to obtain boron–polyol complexes. Results obtained suggest that adding polyols allows NF membranes to be effectively used for boron removal without any considerable membrane fouling.

Polymer-assisted ultrafiltration (PAUF) has been introduced by Smith et al.[82] in which soluble polyalcohol macromolecules were used. As a feed solution, 10 mg/L of boron solution at pH 9.0 have been used and 1 mg/L boron concentration in permeate has been reported. Due to the small-scale dimensions of the membrane modules, PAUF was considered an alternative to boron-specific ion exchange resins for the treatment of water streams containing relatively low boron concentrations. Some researchers have preferred to name the same concept as PEUF involving water-soluble sorbents, which create boron complexations that can be separated by UF based on the mechanism of size exclusion.[4,83–85]

Yurum et al.[83] studied three types of boron sorbents: NMDG-grafted poly(glycidyl methacrylate), iminodipropylene-glycol-grafted poly(glycidyl methacrylate), and a gel form of the latter. It was observed that the NMDG-based polymer performed the best due to the relatively linear branches facilitating the borate complexations. Up to 90% of boron retentions were observed with different pH values.

On the contrary, Zerze et al.[84] used a chelating polymer poly(vinyl amino-N,N'-bis-propane diol) for boron removal. Boron removal efficiency was as high as 96% at pH 9.0 in a total recycle PEUF system using a 10 mg/L boron feed solution. In another study of Zerze et al.,[4] a copolymer, poly(vinyl amino-N,N'-bis-propane diol-co-DADMAC) in three comonomer ratios (5%, 10%, and 15%) was synthesized to be used in PEUF. The highest boron retention observed was 92% from a solution containing 10 mg B/L at a pH of 9.0.

2.4.2   Hybrid Processes Using Boron-Selective Chelating Ion Exchange Resins

Adsorption membrane filtration (AMF) is another method in which water-insoluble sorbents are usually used. The use of boron-selective microspherical resins followed by MF is combined in this technology such that:

1. Boron is sorbed by boron-selective chelating ion exchange resins in water.
2. Saturated boron-selective ion exchange resins are separated from water.
3. Selective ion exchange resins are regenerated.
4. Boron-selective ion exchange resins are recycled.

The system containing membrane filtration and resins can be considered as in the same analogy as fluidized bed systems. However, this configuration does not need high pressure and the membranes can withstand relatively high flux conditions. Submerged membrane units are usually preferred over cross-flow geometry due to the lower energy requirements. Such advantages also allow the use of adsorbents with much smaller size as in powder form rather than in beads geometry. It should also be noted that the sorbents for AMF applications should have smaller diameters, uniform sizes, and a smooth surface morphology to protect the membranes from physical damage.[85]

This application has been further reviewed by Kabay et al.[16] by investigating the treatment of aqueous solution containing 1.7 mg/L of boron. For this, the Dowex XUS boron-selective chelating ion exchange resin has been employed. The same researchers also used Diaion CRB02 resins for boron removal.[49] On the contrary, the particle sizes of the commercially available BSR are too large to be implemented in current AMF systems. Tailoring those resins down to smaller particle size (e.g., 50 μm) is a current issue for researchers.[86] Later, these resins (Dowex and Diaion) were used to remove boron from geothermal water containing relatively high levels of boron concentrations (8–9 mg/L).[49] The main highlight of the results was that when the resin size decreased from 45–125 to 0–20 μm, boron removal increased significantly due to the increased area of the resin. Moreover, when the required resin concentration was 2 g/L (of Dowex XUS 43594.00), boron concentration in the product water was well below the permissible level for drinking water (0.3 mg/L). In addition to those results, Kabay et al.[49] also suggested following three paths to reduce boron further in the submerged systems:

1. Reduce the particle size of the resin.
2. Decrease the delivery rate of the feed.
3. Increase the resin concentration in the suspension.

Bryjak et al.[87] used 20 μm particles of the same Dowex XUS sorbent to remove boron from a typical first-stage RO permeate (containing 2 mg/L of boron). The removal of the trace amount of boron appeared within 2–3 min of sorption and it was enough to use 1 g of the sorbent per 1 L of the feed to keep the boron concentration below 1 mg/L. Wolska and Bryjak[69] further developed polymeric microspheres by using different amounts of vinylbenzyl chloride–styrene–divinyl benzene monomers.

Later on, these microspheres were modified with NMG to make them boron-selective materials with 30 μm particle size and narrow pore size distribution to be employed in an AMF process for boron removal. Elsewhere, the preparations of monodisperse nanoporous NMGD bearing poly(GMA-*co*-EDM) and poly(VBC-*co*-DVB) beads of about 9–10 μm of particle size using "modified seeded polymerization" were reported by Samatya et al. The synthesized monodisperse resin beads were very effective for boron sorption.[71–73]

As was previously stated, a sorption–membrane filtration hybrid method can take an important role in boron removal from both seawater and geothermal water streams.[16] Boron removal from seawater can be performed directly from seawater or a first-stage RO permeate.[53,88] The first systematic simulation and optimization of a sorption-microfiltration hybrid process based on experimental data was performed by Blahušiak and Schlosser.[89] The effect of various process parameters on the performance of a hybrid process was studied to determine the investment and operating costs.[89,90]

Kabay et al.[16] and Guler et al.[91] investigated the possibility of treating seawater RO permeate with sorption-microfiltration hybrid processes. Several process parameters, such as sorbent concentration in feed suspension, exchange rate of saturated sorbent, permeate flux, and membrane type, were investigated in order to find optimum conditions for efficient boron removal. Guler et al.[91] showed the applicability of the ZeeWeed® (ZW-1, GE Zenon) submerged UF system for boron removal with an integrated aeration system that prevents the accumulation of sorbent particles on the membrane surface (Figure 2.17).

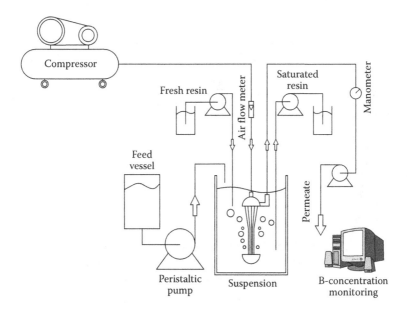

FIGURE 2.17 Sorption–UF membrane hybrid system for boron removal from seawater permeate. (Reprinted from *Journal of Membrane Science*, 375, Guler, E., et al., Integrated solution for boron removal from seawater using RO process and sorption-membrane filtration hybrid method, 249–257, Copyright 2011, with permission from Elsevier.)

For geothermal water streams that usually exhibit higher boron concentrations than that of seawater, similar process parameters were investigated using boron-specific ion exchange resins (Diaion CRB 02 and Dowex XUS) by Kabay et al.[49] Moreover, Yilmaz-Ipek et al.[92] modeled nonequilibrium sorption for boron removal from geothermal water using a sorption–MF hybrid approach. Most recently, Kabay et al.[93] studied a submerged UF membrane module (ZW-1, GE) for boron removal from geothermal water. The recycle performance of that system was also evaluated. As conclusions, it was possible to obtain product water containing well below 1 mg/L boron approximately in 20 min during 10 cycles and this concentration level was kept constant until the end of the experiment (Figure 2.18). During the cycle study, no significant change in the performance of the resin was observed.

The same system was used by Kabay et al.[94] to remove boron from geothermal water RO permeate containing boron at a concentration range of 4.75–5.15 mg/L. The product water had a low level of boron concentration (even lower than 0.3 mg/L), indicating that it could be used as irrigation water (Figure 2.19). In that particular work, it was found that to lower the boron concentration in the geothermal water RO permeate below 1.0 mg/L and keep this concentration constant throughout the experiment, the optimum resin concentration and the speed of resin replacement should be adjusted to 1.5 g/L and 6 mL/min, respectively, while the flow rate of the permeate is kept constant as 10 mL/min. These results are crucial and promising when the quality requirements of the irrigation water are considered such as low salinity and acceptable boron level.

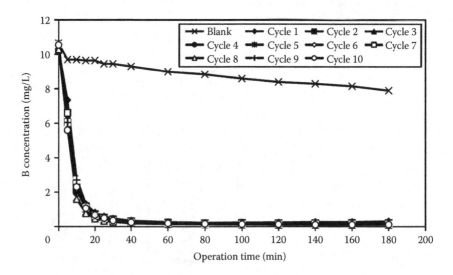

FIGURE 2.18 Boron concentration in the permeate versus time plots for 10 cycles. (Reprinted from *Desalination*, 316, Kabay, N., et al., Coupling ion exchange with ultrafiltration for boron removal from geothermal water-investigation of process parameters and recycle tests, 17–22, Copyright 2013, with permission from Elsevier.)

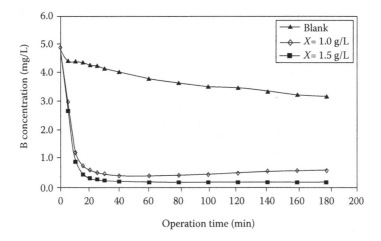

FIGURE 2.19   Effect of resin concentration (X) on boron concentration of the permeate (speed of saturated resin replacement = 6 mL/min). (Reprinted from *Desalination*, 316, Kabay, N., et al., An innovative integrated system for boron removal from geothermal water using RO process and ion exchange-ultrafiltration hybrid method, 1–7, Copyright 2013, with permission from Elsevier.)

2.5   CONCLUSIONS

Although boron is essential as a trace element for plant growth, it can be detrimental at high concentrations. In 2011, the Drinking-Water Quality Committee of the WHO revised the Boron Guideline Value as 2.4 mg/L in drinking water. However, some utilities may still set desalination plants product water limits as low as 0.3–0.5 mg/L bearing in mind the toxic effect of higher concentrations of boron to some plants. Several treatment methods were developed for boron removal from water. Among the proposed methods, an ion exchange method employing boron-selective ion exchange resins is still the most promising one for eliminating boron. However, there are still continuing requests for producing boron-selective chelating ion exchange resins with a higher capacity, fast kinetic, and lower cost for boron removal from water. Recently, a hybrid process integrating sorption using boron-selective chelating ion exchange resins with membrane filtration seemed to be the challenging method for decreasing the concentration of boron in water to permissible level with a lower cost than with a fixed-bed ion exchange process. Indeed, the combination of the efficiency of ion exchange resins with easiness of membrane separations reveals many advantages. The main benefit of the hybrid method over the classical fixed-bed column process is a reduction in consumption of reactants, which means low cost of the process. The hybrid process has the more efficient use of boron-selective chelating ion exchange resins with much faster kinetics of sorption in microparticles. However, the performance of the hybrid system should be checked with a longer period of time in a pilot-scale plant installed in a real media.

ACKNOWLEDGMENTS

The authors thank their coworkers and students cited in the various references of this chapter and other students/coworkers for their great collaborations. N. Kabay thanks TÜBİTAK (ÇAYDAG-104I096; TÜBİTAK-JSPS-214M360), BOREN (project number: 2008-G-0192), Ege University Scientific Research projects Commission (project numbers: 2004-MÜH-026, 2004-BIL-004, 2005-MÜH-033, 2007-MUH-015, 2008-MÜH-029), MEDRC (project number: 04-AS-004), NATO-CLG, BSEC, and NEDO for their financial support and the Dow Chemical Co., Mitsubishi Chemical Co., and Purolite International Co. for sending boron-selective ion exchange resins to let them get experience on boron removal from water by ion exchange and hybrid processes.

REFERENCES

1. Shih, Y. J., Liu, C. H., Lan, W. C., Huang, Y. H. A novel chemical oxo-precipitation (COP) process for efficient remediation of boron wastewater at room temperature. *Chemosphere*. **2014**, *111*, 232–237.
2. Zelmanov, G., Semiat, R. Boron removal from water and its recovery using iron (Fe^{+3}) oxide/hydroxide-based nanoparticles (NanoFe) and NanoFe-impregnated granular activated carbon as adsorbent. *Desalination*. **2014**, *333*, 107–117.
3. Sabarudin, A., Oshita, K., Oshima, M., Motomizu, S. Synthesis of cross-linked chitosan possessing N-methyl-d-glucamine moiety (CCTS-NMDG) for adsorption/concentration of boron in water samples and its accurate measurement by ICP-MS and ICP-AES. *Talanta*. **2005**, *66*, 136–144.
4. Zerze, H., Ozbelge, H. O., Bicak, N., Aydogan, N., Yilmaz, L. Novel boron specific copolymers with quaternary amine segments for efficient boron removal via PEUF. *Desalination*. **2013**, *310*, 169–179.
5. MEDRC Research Project (Project No. 04-AS-004). *Study of the Adsorption-Membrane Filtration (AMF) Hybrid Process for Removal of Boron from Seawater*, Ege University, İzmir, **2007**.
6. Redondo, J., Busch, M., De Witte, J. P. Boron removal from seawater using FILMTEC™ high rejection SWRO membranes. *Desalination*. **2003**, *156*, 229–238.
7. Vengosh, A., Barth, S., Heumann, K. G., Eisenhut, S. Boron isotopic composition of fresh water lakes from Central Europe and possible contamination sources. *Acta Hydrochim. Hydrobiol.* **1999**, *27*, 416–421.
8. Yavuz, E., Güler, E., Sert, G., Arar, Ö., Yüksel, M., Yüksel, Ü., Kitiş, M., Kabay, N. Removal of boron from geothermal water by RO system-I—effect of membrane configuration and applied pressure. *Desalination*. **2013**, *310*, 130–134.
9. Pastor, M. R., Ruiz, A. F., Chillon, M. F., Rico, D. P. Influence of pH in the elimination of boron by means of reverse osmosis. *Desalination*. **2001**, *140*, 145–152.
10. Davis, S. M., Drake, K. D., Maier, K. J. Toxicity of boron to the duckweed, Spirodella polyrrhiza. *Chemosphere*. **2002**, *48*, 615–620.
11. Green Facts. Scientific facts on boron, IPCS, 1998. http://www.greenfacts.org/en/boron/l-2/boron-3.htm#1.
12. Nadav, N. Boron removal from seawater reverse osmosis permeate utilizing selective ion exchange resin. *Desalination*. **1999**, *164*, 131–135.
13. Palencia, M., Vera, M., Rivas, B. L. Modification of ultrafiltration membranes via interpenetrating polymer networks for removal of boron from aqueous solution. *J. Membr. Sci.* **2014**, *466*, 192–199.

14. Boncukcuoğlu, R., Yilmaz, A. E., Kocakerim, M. M., Çopur, M. An empirical model for kinetics of boron removal from boron containing wastewaters by ion exchange in a batch reactor. *Desalination.* **2004**, *160*, 159–166.
15. Sagiv, A., Semiat, R. Analysis of parameters affecting boron permeation through reverse osmosis membranes. *J. Membr. Sci.* **2004**, *243*, 79–87.
16. Kabay, N., Güler, E., Bryjak, M. Boron in seawater and methods for its separation—A review. *Desalination.* **2010**, *261*, 212–217.
17. Yurdakoc, M., Seki, Y., Karahan, S., Yurdakoc, K. Kinetic and thermodynamic studies of boron removal by Siral 5, Siral 40, and Siral 80. *J. Colloid. Interface Sci.* **2005**, *286*, 440–446.
18. Seki, Y., Seyhan S., Yurdakoc, M. Removal of boron from aqueous solution by adsorption on Al_2O_3 based materials using full factorial design. *J. Hazard. Mater.* **2006**, *B138*, 60–66.
19. Polat, H., Vengosh, A., Pankratov, I., Polat, M. A new methodology for removal of boron from water by coal and fly ash. *Desalination.* **2004**, *164*, 173–188.
20. Bouguerra, W., Mnif, A., Hamrouni, B., Dhahbi, M. Boron removal by adsorption onto activated alumina and by reverse osmosis. *Desalination.* **2008**, *223*, 31–37.
21. Tagliabue, M., Reverberi, A. P., Bagatin, R. Boron removal from water: Needs, challenges and perspectives. *J. Clean. Prod.* **2014**, *77*, 56–64.
22. Demey, H., Vincent, T., Ruiz, M., Nogueras, M., Sastre, A. M., Guibal, E. Boron recovery from seawater with a new low-cost adsorbent material. *Chem. Eng. J.* **2014**, *254*, 463–471.
23. Itakura, T., Sasai, R., Itoh, H. Precipitation recovery of boron from wastewater by hydrothermal mineralization. *Water Res.* **2005**, *39*, 2543–2548.
24. Irawan, C., Kuo, Y. L., Liu, J. C. Treatment of boron-containing optoelectronic wastewater by precipitation process. *Desalination.* **2011**, *280*, 146–151.
25. Yilmaz, A. E., Boncukcuoğlu, R., Kocakerim, M. M. A quantitative comparison between electrocoagulation and chemical coagulation for boron removal from boron-containing solution. *J. Hazard. Mater.* **2007**, *149*, 475–481.
26. Yılmaz, A. E., Boncukcuoğlu, R., Kocakerim, M. M., Kocadağistan, E. An empirical model for kinetics of boron removal from boron containing wastewaters by the electro-coagulation method in a batch reactor. *Desalination.* **2008**, *230*, 288–297.
27. Yilmaz, A. E., Boncukcuoğlu, R., Kocakerim, M. M., Yilmaz, M. T., Paluluoglu, C. Boron removal from geothermal waters by electrocoagulation. *J. Hazard. Mater.* **2008**, *153*, 146–151.
28. Pieruz, G., Grassia, E., Dryfe, R. A. W. Boron removal from produced water by facilitated ion transfer. *Desalination.* **2004**, *167*, 417–423.
29. Karakaplan, M., Tural, S., Sunkur, M., Hosgoren, H. Effect of 1,3-diol structure on the distribution of boron between $CHCl_3$ and aqueous phase. *Sep. Sci. Technol.* **2003**, *38*(8), 1721–1732.
30. Fujita, Y., Hata, T., Nakamaru, M., Iyo, T., Yoshino, T., Shimamura, T. A study of boron adsorption onto activated sludge. *Bioresour. Technol.* **2005**, *96*, 1350–1356.
31. Xu, L., Liu, Y., Hu, H., Wu, Z., Chen, Q. Synthesis, characterization and application of a novel silica based adsorbent for boron removal. *Desalination.* **2012**, *294*, 1–7.
32. Demey, H., Vincent, T., Ruiz, M., Sastre, A. M., Guibal, E. Development of a new chitosan/Ni(OH)$_2$-based sorbent for boron removal. *Chem. Eng. J.* **2014**, *244*, 576–586.
33. Prats, D., Chillon-Arias, M. F., Rodriguez-Pastor, M. Analysis of the influence of pH and pressure on the elimination of boron in reverse osmosis. *Desalination.* **2000**, *128*, 269–273.
34. Bick, A., Oron, G. Post-treatment design of seawater reverse osmosis plants boron removal technology selection for potable water production and environmental control *Desalination.* **2005**, *178*, 233–246.

35. Ozturk, N., Kavak, D., Köse, T. E. Boron removal from aqueous solution by reverse osmosis. *Desalination.* **2008**, *223*, 1–9.
36. Melnik, L., Vysotskaja, O., Kornilovich, B. Boron behaviour during desalination of sea and underground water by electrodialysis. *Desalination.* **1999**, *124*, 125–130.
37. Kabay, N., Arar, O., Acar, F., Ghazal, A., Yuksel, U., Yuksel, M. Removal of boron from water by electrodialysis: Effect of feed characteristics and interfering ions. *Desalination.* **2008**, *223*, 63–72.
38. Yavuz, E., Arar, Ö., Yüksel, M., Yüksel, Ü., Kabay, N. Removal of boron from geothermal water by RO system-II-effect of pH. *Desalination.* **2013**, *310*, 135–139.
39. Yavuz, E., Arar, Ö., Yüksel, Ü., Yüksel, M., Kabay, N. Removal of boron from geothermal water by RO system-III-utilization of SWRO system. *Desalination.* **2013**, *310*, 140–144.
40. Guler, E., Kabay, N., Yüksel, M., Yavuz, E., Yüksel, Ü. A comparative study for boron removal from seawater by two types of polyamide thin film composite SWRO membranes. *Desalination.* **2011**, *273*, 81–84.
41. Arar, Ö., Yüksel, Ü., Kabay, N., Yüksel, M. Application of electrodeionization (EDI) for removal of boron and silica from reverse osmosis (RO) permeate of geothermal water. *Desalination.* **2013**, *310*, 25–33.
42. Öner, Ş. G., Kabay, N., Güler, E., Kitiş, M., Yüksel, M. A comparative study for the removal of boron and silica from geothermal water by cross-flow flat sheet reverse osmosis method. *Desalination.* **2011**, *283*, 10–15.
43. Hilal, N., Kim, G. J., Somerfield, C. Boron removal from saline water: A comprehensive review. *Desalination.* **2011**, *273*, 23–35.
44. Guler, E., Ozakdag, D., Arda, M., Yuksel, M., Kabay, N. Effect of temperature on seawater desalination-water quality analyses for desalinated seawater for its use as drinking and irrigation water. *Environ. Geochem. Health.* **2010**, *32*(4), 335–339.
45. Jacob, C. Seawater desalination: Boron removal by ion exchange technology. *Desalination.* **2007**, *205*, 47–52.
46. Simonnot, M. O., Castel, C., Nicolai, M., Rosin, C., Sardin, M., Jauffret, H. Boron removal from drinking water with a boron selective resin: Is the treatment really selective? *Water Res.* **2000**, *34*, 109–116.
47. Glueckstern, P., Priel, M. Boron removal in brackish water desalination systems. *Desalination.* **2007**, *205*, 178–184.
48. Arias, M. F. C., Valerio Bru, L., Rico, D. P., Varo Galvan, P. Comparison of ion exchange resins used in reduction of boron in desalinated water for human consumption. *Desalination.* **2011**, *278*, 244–249.
49. Kabay, N., Yilmaz Ipek, I., Soroko, I., Makowski, M., Kirmizisakal, O., Yag, S., Bryjak, M., Yuksel, M. Removal of boron from Balcova geothermal water by ion exchange–microfiltration hybrid process. *Desalination.* **2009**, *241*, 167–173.
50. Yilmaz, I., Kabay, N., Bryjak, M., Yuksel, M., Wolska, J., Koltuniewicz, A. A submerged membrane–ion-exchange hybrid process for boron removal. *Desalination.* **2006**, *198*, 310–315.
51. Kabay, N., Yilmaz, I., Bryjak, M., Yuksel, M. Removal of boron from aqueous solutions by a hybrid ion exchange–membrane process. *Desalination.* **2006**, *198*, 158–165.
52. Kunin, R., Preuss, A. F. Characterization of a boron-specific resin. *I & EC Prod. Res. Dev.* **1964**, *3*(4), 304–306.
53. Kabay, N., Bryjak, M., Schlosser, S., Kitis, M., Avlonitis, S., Matejka, Z., Al-Mutaz, I., Yuksel, M. Adsorption-membrane filtration (AMF) hybrid process for boron removal from seawater: An overview. *Desalination.* **2008**, *223*, 38–48.
54. Busch, M., Marston, C., Prabakaran, C. A boron selective resin for seawater desalination, *Proceedings of European Desalination Society Conference on Desalination and Environment*, Santa Margherita, Italy, 2005.

55. Kabay, N., Yılmaz, I., Yamac, S., Samatya, S., Yuksel, M., Yuksel, U., Arda, M., Sağlam, M., Iwanaga, T., Hirowatari, K. Removal and recovery of boron from geothermal wastewater by selective ion exchange resins. I. Laboratory tests. *React. Func. Polym.* **2004**, *60*, 163–170.

56. Kabay, N., Yılmaz, I., Yamac, S., Yuksel, M., Yuksel, U., Yildirim, N., Aydogdu, O., Iwanaga, T., Hirowatari, K. Removal and recovery of boron from geothermal wastewater by selective ion-exchange resins—II. Field tests. *Desalination.* **2004**, *167*, 427–438.

57. Yilmaz, I., Kabay, N., Yuksel, M., Holdich, R., Bryjak, M. Effect of ionic strength of solution on boron mass transfer by ion exchange separation. *Sep. Sci. Technol.* **2007**, *42*, 1013–1029.

58. Nishihama, S., Sumiyoshi, Y., Ookubo, T., Yoshizuka, K. Adsorption of boron using glucamine-based chelate adsorbents. *Desalination.* **2013**, *310*, 81–86.

59. Li, X., Liu, R., Wua, S., Liu, J., Cai, S., Chen, D. Efficient removal of boron acid by *N*-methyl-D-glucamine functionalized silica–polyallylamine composites and its adsorption mechanism. *J. Colloid. Interface Sci.* **2011**, *361*, 232–237.

60. Wei, Y. T., Zheng, Y. M., Chen, J. P. Design and fabrication of an innovative and environmental friendly adsorbent for boron removal. *Water Res.* **2011**, *45*, 2297–2305.

61. Thakura, N., Kumara, S. A., Shinde, R. N., Pandey, A. K., Kumara, S. D., Reddy, A. V. R. Extractive fixed-site polymer sorbent for selective boron removal from natural water. *J. Hazard. Mater.* **2013**, *260*, 1023–1031.

62. Mehltretter, C. L., Weakley, F. B., Wilham, C. A. Boron-selective ion exchange resins containing D-glucitylamino radicals. *Prod. Res. Dev.* **1967**, *6*(3), 145–147.

63. Maeda, H., Egawa, H. Preparation of chelating resins selective to boric acid by functionalization of macroporous poly(glycidyl methacrylate) with 2-amino-2-hydroxymethyl-1, 3-propanediol. *Sep. Sci. Technol.* **1995**, *30*(18), 3545–3554.

64. Ristic, M., Rajakoviç, Lj. Boron removal by anion exchangers impregnated with citric and tartaric acids. *Sep. Sci. Technol.* **1996**, *31*(20), 2805–2814.

65. Inukai, Y., Tanaka, Y., Matsuda, T., Mihara, N., Yamada, K., Nambu N., Itoh O., Doi T., Kaida Y., Yasuda S. Removal of boron (III) by *N*-methylglucamine-type cellulose derivatives with higher adsorption rate. *Anal. Chim. Acta.* **2004**, *511*, 261–265.

66. Yilmaz-Ipek, I., Koseoglu, P., Yuksel, U., Yasar, N., Yolseven, G., Yuksel, M., Kabay, N. Separation of boron from geothermal water using a boron selective macroporous weak base anion exchange resin. *Sep. Sci. Techol.* **2010**, *45*, 809–813.

67. Santander, P., Rivas, B. L., Urbano, B. F., Yılmaz İpek, İ., Özkula, G., Arda, M., Yüksel, M., Bryjak, M., Kozlecki, T., Kabay, N. Removal of boron from geothermal water by a novel boron selective resin. *Desalination.* **2013**, *310*, 102–108.

68. Wolska, J., Bryjak, M., Kabay, N. Polymeric microspheres with *N*-methyl-D-glucamine ligands for boron removal from water solution by adsorption–membrane filtration process. *Environ. Geochem. Health.* **2010**, *32*, 349–352.

69. Wolska, J., Bryjak, M. Preparation of polymeric microspheres for removal of boron by means of sorption-membrane filtration hybrid. *Desalination.* **2011**, *283*(1), 193–197.

70. Samatya, S., Orhan, E., Kabay, N., Tuncel, A. Comparative boron removal performance of monodisperse-porous particles with molecular brushes via "click chemistry" and direct coupling. *Colloids Surf. A.* **2010**, *372*, 102–106.

71. Samatya, S., Kabay, N., Tuncel, A. A hydrophilic matrix for boron isolation: Monodisperse-porous poly(glycidyl methacrylate-co-ethylene dimethacrylate) particles carrying diol functionality. *React. Funct. Polym.* **2010**, *70*, 555–562.

72. Samatya, S., Tuncel, A., Kabay, N. Boron removal from geothermal water by a novel monodisperse porous poly(GMA-*co*-EDM) resin containing *N*-methyl-D-glucamine functional group. *Solvent Extr. Ion Exch.* **2012**, *30*(4), 341–349.

73. Samatya, S., Kabay, N., Tuncel, A. Monodisperse porous N-methyl-D-glucamine functionalized poly(vinylbenzyl chloride-co-divinylbenzene)beads as boron selective sorbent. *J. Appl. Polym. Sci.* **2012**, *126*(4), 1475–1483.

74. Bilgin Simsek, E., Beker, U., Senkal, B. F. Predicting the dynamics and performance of selective polymeric resins in a fixed bed system for boron removal. *Desalination.* **2014**, *349*, 39–50.

75. Wang, B., Lin, H., Guo, X., Bai, P. Boron removal using chelating resins with pyrocatechol functional groups. *Desalination.* **2014**, *347*, 138–143.

76. Yilmaz-Ipek, I. *Boron removal from geothermal water by ion exchange-membrane hybrid process.* PhD Thesis, Ege University, 2009.

77. Bekci, Z., Seki, Y., Yurdakoc, M. K. A study of equilibrium and FTIR, SEM/EDS analysis of trimethoprim adsorption onto K10. *J. Mol. Struct.* **2007**, *827*, 67–74.

78. Geffen, N., Semiat, R., Eisen, M. S., Balazs, Y., Katz, I., Desoretz, C. G. Boron removal from water by complexation to polyol compounds. *J. Membr. Sci.* **2006**, *286*, 45–51.

79. Dydo, P., Turek, M. The concept for an ED–RO integrated system for boron removal with simultaneous boron recovery in the form of boric acid. *Desalination.* **2014**, *342*, 35–42.

80. Dydo, P., Turek, M., Milewski, A. Removal of boric acid, monoborate and boron complexes with polyols by reverse osmosis membranes. *Desalination.* **2014**, *334*, 39–45.

81. Tu, K. L., Chivas, A. R., Nghiem, L. D. Enhanced boron rejection by NF/RO membranes by complexation with polyols: Measurement and mechanisms. *Desalination.* **2013**, *310*, 115–121.

82. Smith, B. M., Todd, P., Bowman, C. Hyperbranched chelating polymers for the polymer-assisted ultrafiltration of boric acid. *Sep. Sci. Technol.* **1999**, *34*(10), 1925–1945.

83. Yurum, A., Taralp, A., Bicak, N., Ozbelge, H. O., Yilmaz, L. High performance ligands for the removal of aqueous boron species by continuous polymer enhanced ultrafiltration. *Desalination.* **2013**, *320*, 33–39.

84. Zerze, H., Karagoz, B., Ozbelge, H. O., Bicak, N., Aydogan, N., Yilmaz, L. Imino-bis-propane diol functional polymer for efficient boron removal from aqueous solutions via continuous PEUF process. *Desalination.* **2013**, *310*, 158–168.

85. Wang, B., Guo, X., Bai, P. Removal technology of boron dissolved in aqueous solutions—A review. *Colloid Surf. A.* **2014**, *444*, 338–344.

86. Wolska, J., Bryjak, M. Methods for boron removal from aqueous solutions—A review. *Desalination.* **2013**, *310*, 18–24.

87. Bryjak, M., Wolska, J., Soroko, I., Kabay, N. Adsorption-membrane filtration process in boron removal from first stage seawater RO permeate. *Desalination.* **2009**, *241*, 127–132.

88. Bryjak, M., Wolska, J., Kabay, N. Removal of boron from seawater by adsorption–membrane hybrid process: Implementation and challenges. *Desalination.* **2008**, *223*, 57–62.

89. Blahušiak, M., Schlosser, Š. Simulation of the adsorption—Microfiltration process for boron removal from RO permeate. *Desalination.* **2009**, *241*, 156–166.

90. Blahušiak, M., Onderková, B., Schlosser, Š., Annus, J. Microfiltration of microparticulate boron adsorbent suspensions in submerged hollow fibre and capillary modules. *Desalination.* **2009**, *241*, 138–147.

91. Guler, E., Kabay, N., Yuksel, M., Yiğit, N. Ö., Kitiş, M., Bryjak, M. Integrated solution for boron removal from seawater using RO process and sorption-membrane filtration hybrid method. *J. Membr. Sci.* **2011**, *375*, 249–257.

92. Yilmaz-Ipek, I., Kabay, N., Ozdural, A. Non-equilibrium sorption modeling for boron removal from geothermal water using sorption–microfiltration hybrid method. *Chem. Eng. Process.* **2011**, *50*, 599–607.

93. Kabay, N., Yuksel, M., Koseoglu, P., Yapıcı, D., Yuksel, U. Coupling ion exchange with ultrafiltration for boron removal from geothermal water-investigation of process parameters and recycle tests. *Desalination.* **2013**, *316*, 17–22.
94. Kabay, N., Koseoglu, P., Yavuz, E., Yuksel, U., Yuksel, M. An innovative integrated system for boron removal from geothermal water using RO process and ion exchange-ultrafiltration hybrid method. *Desalination.* **2013**, *316*, 1–7.

3 Greener and Sustainable Approach for the Synthesis of Commercially Important Epoxide Building Blocks Using Polymer-Supported Mo(VI) Complexes as Catalysts

Misbahu Ladan Mohammed and Basudeb Saha

CONTENTS

The growing concern for the environment, increasingly stringent standards for the release of chemicals into the environment, and economic competitiveness have prompted extensive efforts to improve chemical synthesis and manufacturing methods as well as the development of new synthetic methodologies that minimize or completely eliminate pollutants. As a consequence, more and more attention has been focused on the use of safer chemicals through proper design of clean processes and products. Epoxides are key raw materials or intermediates in organic synthesis, particularly for the functionalization of substrates and the production of a wide variety of chemicals such as pharmaceuticals, plastics, paints, perfumes, food additives, and adhesives. The conventional methods for the industrial production of epoxides employ either stoichiometric peracids or chlorohydrin as an oxygen source. However, both methods have serious environmental impact as the former produces an equivalent amount of acid waste, while the latter yields chlorinated by-products and calcium chloride waste. Hence, a greener and efficient route for catalytic epoxidation that could improve manufacturing efficiency by reducing operational cost and minimizing waste products is highly desired.

In this chapter, a greener alkene epoxidation process using a molybdenum (Mo)-based heterogeneous catalyst and *tert*-butyl hydroperoxide (TBHP) as an oxidant has been presented. A polystyrene 2-(aminomethyl)pyridine–supported molybdenum(VI) complex, that is, Ps.AMP.Mo and a polybenzimidazole (PBI)-supported Mo(VI) complex, i.e., PBI.Mo, has been successfully prepared and characterized. The catalytic activities of the polymer-supported Mo(VI) complexes have been evaluated for epoxidation of 1-hexene and 4-vinyl-1-cyclohexene (4-VCH) in a batch reactor. Experiments have been carried out to study the effect of reaction temperature, feed molar ratio (FMR) of alkene to TBHP, and catalyst loading on the yield of epoxide for optimization of reaction conditions in a batch reactor. The long-term stability of the polymer-supported Mo(VI) catalysts has been evaluated by recycling the catalysts several times in batch experiments using conditions that form the basis for continuous epoxidation studies. The extent of Mo leaching from each polymer-supported catalyst has been investigated by isolating any residue from reaction supernatant studies after removal of heterogeneous catalyst and using the residue as a potential catalyst for epoxidation. The efficiency of the Ps.AMP.Mo catalyst has been assessed for continuous epoxidation of 1-hexene and 4-VCH with TBHP as an oxidant using a FlowSyn reactor by studying the effect of reaction temperature, FMR of alkene to TBHP, and feedflow rate on the conversion of TBHP and the yield of epoxide. The catalysts were found to be active and selective for batch and continuous epoxidation of alkenes using TBHP as an oxidant. The continuous epoxidation in a FlowSyn reactor has shown considerable time savings, high reproducibility and selectivity, along with remarkable improvement in catalyst stability compared to the reactions carried out in a batch reactor.

3.1 INTRODUCTION

Chemical industries are continuously facing increasing challenges of regulatory requirements and rising costs of manufacturing intermediates. Hence, a greener and efficient route for chemical synthesis that could improve manufacturing efficiency by reducing operational cost and minimizing waste products is highly desired. Alkene epoxidation is the addition of oxygen to alkene to yield epoxide. The reaction is very useful in organic synthesis as the resultant epoxide is a highly reactive compound that is used as a raw material or intermediate in the production of commercially important products for flavors, fragrances, paints, and pharmaceuticals.[1–9]

The conventional epoxidation methods in the fine chemical industries employ either stoichiometric peracids such as peracetic acid and m-chloroperbenzoic acid[10] or chlorohydrin[11] as oxidizing reagents in liquid-phase batch reactions. However, such processes are not environmentally benign as the former produces an equivalent amount of acid waste, while the latter yields chlorinated by-products and calcium chloride waste. In recent years, more and more attention has been focused on developing greener and more efficient epoxidation processes, employing environmentally benign oxidants such as tert-butyl hydroperoxide (TBHP) since it is atom efficient and safer to handle.[12,13] A notable industrial implementation of alkene epoxidation with TBHP was the Halcon process described by Kollar,[14] which employs homogeneous molybdenum (VI) as a catalyst for liquid-phase epoxidation of propylene to propylene oxide (PO). However, homogeneous-catalyzed epoxidation processes are not economically viable for industrial applications due to major requirements in terms of work-up, product isolation, and purification procedures. Therefore, researchers have focused on developing stable heterogeneous catalysts for epoxidation by immobilization of catalytically active metal species on organic or inorganic materials.[15–17] Polymers have gained attention as suitable support for transition metal catalysts as they are inert, nontoxic, insoluble, and often recyclable.[18] A number of polymer-supported molybdenum complexes have been prepared and used as catalysts for batch alkene epoxidation with TBHP as an oxidant and have shown good catalytic activity and product selectivity.[19–31] However, despite numerous published works on polymer-supported Mo(VI) catalyzed alkene epoxidation with TBHP, there appears to have been no significant efforts to move the chemistry on from a small-scale laboratory batch reaction to a continuous flow process.

In this chapter, we discuss the preparation and characterization of an efficient and selective polystyrene 2-(aminomethyl)pyridine–supported molybdenum complex (Ps.AMP.Mo) and a polybenzimidazole-supported molybdenum complex (PBI.Mo), which have been used as catalysts for epoxidation of 1-hexene and 4-VCH using TBHP as an oxidant. The process is considered to be clean as (i) it employs efficient and selective heterogeneous catalysts; (ii) it is solvent-less; (iii) it uses a benign oxidant (TBHP), which becomes active only on contact with the catalyst; and (v) it is atom efficient and the alcohol by-product itself is an important chemical feedstock.

An extensive assessment of the catalytic activity, stability, and reusability of Ps.AMP.Mo and PBI.Mo catalysts has been conducted in a classical batch reactor. Experiments have been carried out to study the effect of reaction temperature, FMR of alkene to TBHP, and catalyst loading on the yield of 1,2-epoxyhexane and 4-vinyl-1-cyclohexane 1,2-epoxide (i.e., 4-VCH 1,2-epoxide) to optimize the reaction conditions in a batch reactor. A detailed evaluation of molybdenum (Mo) leaching from a polymer-supported catalyst has been investigated by isolating any residue from reaction supernatant solutions and then using these residues as potential catalysts in epoxidation reactions. Furthermore, the efficiency of the heterogeneous catalysts for continuous epoxidation studies has been assessed using a FlowSyn reactor by studying the effect of reaction temperature, FMR of alkene to TBHP, and feedflow rate on the conversion of the oxidant and the yield of corresponding epoxide. The continuous flow epoxidation using a FlowSyn reactor has shown a considerable time savings, high reproducibility and selectivity, along with remarkable improvements in catalyst stability compared to reactions carried out in a batch reactor.

3.1.1 Applications of Epoxides

A wide range of epoxides are produced on a large scale due to their high global demand either as end products or as building blocks in organic synthesis. On the contrary, a number of epoxides are produced on a small scale for specialized, but important applications. The scale of epoxides production ranges from millions of tons per year to a few grams per year depending on requirements and usage.

Over 85% of worldwide consumption of ethylene oxide (EO) is in the production of ethylene glycol, which is widely employed in the manufacture of products such as pharmaceuticals, textiles, automobiles, and detergents.[32] Additionally, EO is used as a sterilant for spices, cosmetics, and medical equipment, and as a fumigant in certain agricultural products.[33] PO is employed in the production of polyurethane polyols, which are used largely for the production of polyurethane foams and plastics. Other applications of PO are found in the production of cosmetics, drugs, and plasticizers as well as in the manufacture of unsaturated polyester resins used in the textile and construction industries.[34] Epichlorohydrin is an important epoxide that is employed as a building block in the manufacture of epoxy resins, glycerols, plastics, and elastomers. One of the largest worldwide consumption of epichlorohydrin is in the production of epoxy resins such as aryl glycidyl ethers,[35] which have several commercial and industrial applications.[36]

Hydroxyethers are useful solvents in the manufacture of lotions, ointments, and creams for pharmaceutical and cosmetic applications. They serve as oil-soluble bases in which a number of lipid-soluble solids can be dissolved efficiently. Hydroxyethers are prepared by acid or alkaline-catalyzed reaction of aliphatic alcohols with terminal epoxides, that is, having 6–18 carbon atoms including 1,2-epoxyhexane, 1,2-epoxyoctane, and 1,2-epoxydecane or their *iso*-derivatives.[37] Epoxides of terminal and internal alkenes are also used as resin modifiers, reactive diluents for epoxy resins, stabilizers for halogen hydrocarbons, and as coating materials.[38] Furthermore, epoxides of cyclic alkenes such as 1,2-epoxycyclohexane are an intermediate for the production of perfumes, pharmaceuticals, and dyestuffs, while 4-VCH 1,2-epoxide is used in the production of coatings and adhesive coupling agents as well as electronic chips encapsulants.

The terpene α-pinene can be epoxidized to obtain a valuable α-pinene oxide. One of the most important industrial reactions of α-pinene oxide is the one-step synthesis of α-campholenic alcohol, also known as naturanol, which is a key ingredient used in the perfumery and food industries due to its natural and sweet berry-like fragrance.[39] Additionally, α-pinene oxide has found its use in the manufacture of some vital unsaturated alcohols that are employed as a sandalwood scent in the fragrance industry. Limonene, which is the main component of orange essential oils, is an abundant, cheap, monoterpene that accumulates in bulk quantities as a by-product of the fruit juice industry.[40,41] Oxidation of limonene mostly yields 1,2-limonene oxide, which is an intermediate in the production of fragrances, perfumes, food additives, and pharmaceuticals.[38]

3.1.2 POLYMERIC SUPPORTS FOR THE IMMOBILIZATION OF CATALYST

Polymer-supported catalysts are made by immobilizing a robust polymer support with an active species either by forming chemical bonds or through physical interactions such as hydrogen bonding or donor–acceptor interactions. Suspension polymerization is one of the most popular and effective methods of synthesizing polymers in spherical or bead form. The polymerization reaction is of two types depending on the nature of the monomers. For instance, inverse suspension polymerization is employed for hydrophilic monomers such as acrylamide using hydrocarbon or chlorinated hydrocarbon as the bulk liquid phase in the reaction.[42] However, water is used as the bulk liquid phase in suspension polymerization involving hydrophobic monomers such as styrene. The polymer resins intended for use as support are normally cross-linked using a bi-functional co-monomer such as divinylbenzene (DVB) to form an infinite network.[43] The uniform shape and sizes of the polymer beads prepared by the suspension polymerization method depend on the shape of the reactor, impeller diameter, stirring speed, porogen, and other reaction conditions.[44] Basically, there are two factors that are responsible for the internal porous structure of polymer beads. These include the amount of crosslinking present and the type of organic solvent or porogen incorporated into the polymer resin. The former determines the level of swelling of the polymer, while the latter creates the pores and influences the pore size, pore volume, and surface area of the polymer particles.[44,45]

Cross-linked polystyrene-based polymers and PBI polymer resins have been used as supports for Mo(VI) complexes used as catalysts in this study. The cross-linked resin beads are commonly used as catalyst supports due to their high porosity, large surface area, and robust spherical particles with uniform size distribution.[43] However, PBI resins are a well-known class of polymers due to their high degree of thermal stability and chemical resistance and have found applications in a wide variety of uses including chromatographic processes (i.e., ion exchange, purifications, and separations), fuel cells, electronics, aerospace, as well as a replacement for asbestos in high-temperature applications such as conveyor belts, plastic composites, and gloves.[46–50]

3.1.3 TBHP AS A TERMINAL OXIDANT FOR ALKENE EPOXIDATION

Epoxidation with alkyl hydroperoxides has received considerable scientific attention in recent times since the reagents are readily available and inexpensive. The commonly employed alkyl hydroperoxides for epoxidation are TBHP, cumene hydroperoxide, and ethyl benzyl hydroperoxide. There are numerous advantages in using TBHP as an oxidant for epoxidation, including high thermal conductivity, good solubility in polar solvents, and neutral pH.[1] In addition, the oxidant is non-corrosive, safer to handle, and yields *tert*-butanol as a by-product, which can be separated easily from the reaction mixture by distillation. Therefore, TBHP was chosen as the terminal oxidant for all the epoxidation experiments reported in this chapter. The commercially available TBHP is in the form of 70% (w/w) solution in 30% water, which serves as a stabilizer. However, the water content could inhibit the epoxidation reaction by causing epoxide openings and the formation of diol by-products.[4] As a result, the oxidant is usually rendered anhydrous in toluene, benzene, or dichloroethane using phase separation and an azeotropic distillation technique.[51]

3.2 CATALYST PREPARATION AND CHARACTERIZATION

The heterogeneous catalysts employed in the present study, that is, a Ps.AMP.Mo and a PBI.Mo, were prepared by immobilization of molybdenum metal species derived from molybdenyl acetylacetonate ($MoO_2(acac)_2$) on two types of polymers, namely a polystyrene 2-aminomethyl(pyridine) (Ps.AMP) and a PBI. A ligand known as 2-(aminomethyl)pyridine (AMP) was first attached to the polystyrene-based resin to act as the site for coordination of the metal center. However, no separate ligand was introduced onto the PBI resin since the benzimidazole residue of the backbone acts as the coordination point for the Mo(VI) center.[52] The catalysts have been characterized extensively to determine their molecular structures, morphological, and physicochemical properties.

3.2.1 PREPARATION OF POLYSTYRENE 2-(AMINOMETHYL)PYRIDINE– SUPPORTED MO(VI) COMPLEX (PS.AMP.MO)

The procedure for preparing Ps.AMP.Mo catalyst involves three main stages, which are described in the following sections.

3.2.1.1 Synthesis of Poly(divinylbenzene-*co*-vinylbenzyl chloride-*co*-styrene) Resin (Ps.VBC)

The polystyrene-based resin used for immobilizing Mo was prepared by a suspension polymerization method. The aqueous continuous phase for the reaction was prepared by dissolving 7.5 g of polyvinyl alcohol (PVOH) in water at ~373 K, followed by the addition of 33 g of sodium chloride. The mixture was stirred until the solids dissolved completely. The solution was allowed to cool down to room temperature and then more water was added until the volume was brought to 1 L, to give a 0.75% PVOH and 3.3% sodium chloride solution. The resulting solution (634 mL) was added to a 1 L parallel-sided, jacketed glass-baffled reactor equipped with a condenser, double impeller, and mechanical stirrer. The volume ratio of organic to the aqueous phase was 1:20.

A flash column chromatography was used to purify each of the organic monomers (DVB, vinylbenzyl chloride [VBC], and styrene) using silica gel and nitrogen gas. The organic phase for the synthesis of 15 g of the resin (Ps.VBC) was prepared by stirring the mixture of DVB (1.95 mL, 12%, 1.8 g), VBC (3.5 mL, 25%, 3.75 g), styrene (10.39 mL, 63%, 9.45 g), 0.15 g of 2,2-azobis isobutyronitrile (AIBN) (equivalent to 1% by weight of the co-monomers), and 2-ethylhexanol (15.84 mL, 1/1 by volume ratio of the co-monomers, 17.03 g) as a porogen in a 0.25 L conical flask. The resulting organic solution was added to the aqueous phase in a 1 L reactor and nitrogen gas was bubbled through the stirred mixture for 30 min before starting the reaction.

The suspension polymerization reaction was carried out under nitrogen atmosphere at 353 K for a period of 6 h, with the stirrer speed set at 500 rpm. After the reaction was completed, the resulting beads were filtered off and washed exhaustively with distilled water. A sonic bath was used to remove other impurities such as traces of NaCl and PVOH that remained on the beads. Finally, the beads were washed successively with methanol and acetone, and then with acetone only. The beads were dried in a vacuum oven at 313 K. The reaction for the synthesis of Ps.VBC resin is illustrated in Figure 3.1.

3.2.1.2 Synthesis of Polystyrene 2-(aminomethyl)pyridine (Ps.AMP) Beads

The synthesis of Ps.AMP resin involves a nucleophilic substitution of the $-CH_2Cl$ functional group of the chloromethylated polystyrene resin (Ps.VBC) by an amino group of AMP. The amination reaction was carried out with an excess of AMP, using a 1:4 mole ratio of Ps.VBC to AMP. For instance, Ps.VBC (35 g, 0.07315 mol

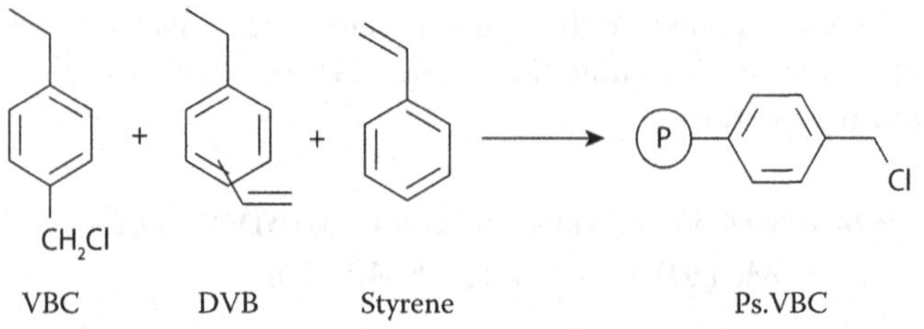

FIGURE 3.1 Synthesis of poly(divinylbenzene-*co*-vinylbenzyl chloride-*co*-styrene) (Ps.VBC) resin.

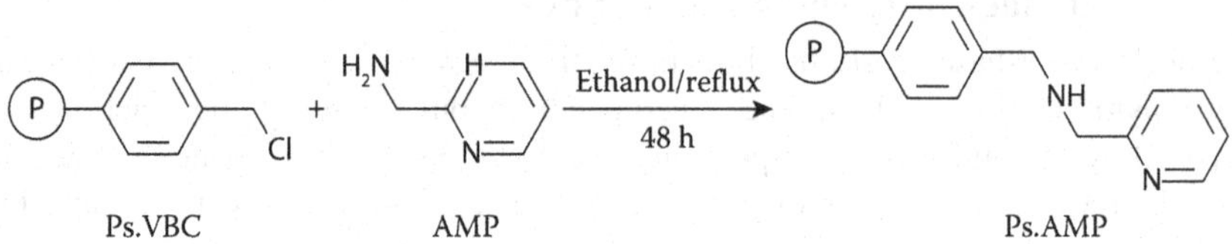

FIGURE 3.2 Synthesis of polystyrene 2-(aminomethyl)pyridine (Ps.AMP) beads.

of Cl) was refluxed with AMP (31.643 g, 30.4 mL, 0.2926 mol) in ~500 mL of ethanol for 48 h, with stirring at ~150 rpm using an overhead mechanical device. The beads were filtered off at the end of the reaction, washed with acetone/water, and then stirred gently overnight in pyridine. The beads were again filtered off and washed successively with water, water/methanol, and acetone, and finally dried in a vacuum oven at 313 K. The reaction for the synthesis of Ps.AMP is shown in Figure 3.2.

3.2.1.3   Loading of Mo(VI) Complex onto Ps.AMP Resin

Ps.AMP resin (17.5 g) was refluxed with an excess of molybdenyl acetylacetonate $(MoO_2(acac)_2)$ (20.77 g) in anhydrous toluene for a period of 4 days. The reaction is illustrated in Figure 3.3. The ratio of $MoO_2(acac)_2$ to functional ligand used was 1:2. The reaction was carried out in a 0.25 L reactor at ~378 K (i.e., close to the boiling point of toluene) and stirred gently with an overhead mechanical device at ~150 rpm. The particles changed from brown to blue during the reaction. The catalyst particles were filtered off at the end of the reaction and extracted exhaustively with acetone to remove the excess $MoO_2(acac)_2$. The dark-blue color for the washings gradually disappeared upon repeated introduction of fresh solvent until the solution became colorless. Finally, the Ps.AMP.Mo catalyst particles were collected and dried in a vacuum oven at 313 K.

3.2.2   Preparation of Polybenzimidazole-Supported Mo(VI) Complex (PBI.Mo)

The wet PBI resin beads were supplied by Celanese Corporation, USA, and were pretreated by stirring in 1 M NaOH solution overnight. The polymer beads were then

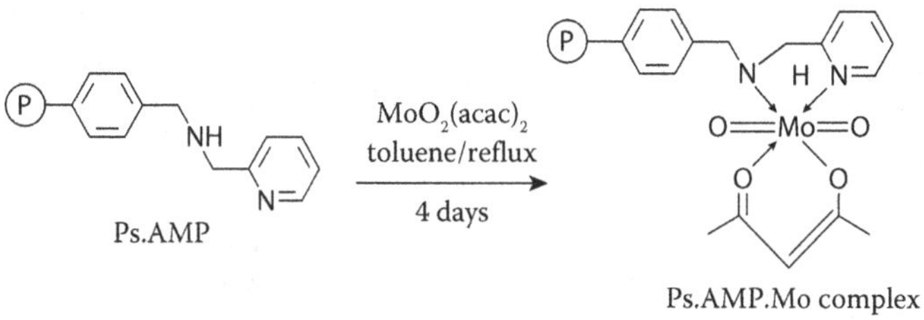

FIGURE 3.3 Loading of Mo(VI) into polystyrene 2-(aminomethyl)pyridine (Ps.AMP) beads to produce polystyrene 2-(aminomethyl)pyridine-supported Mo(VI) (Ps.AMP.Mo) complex.

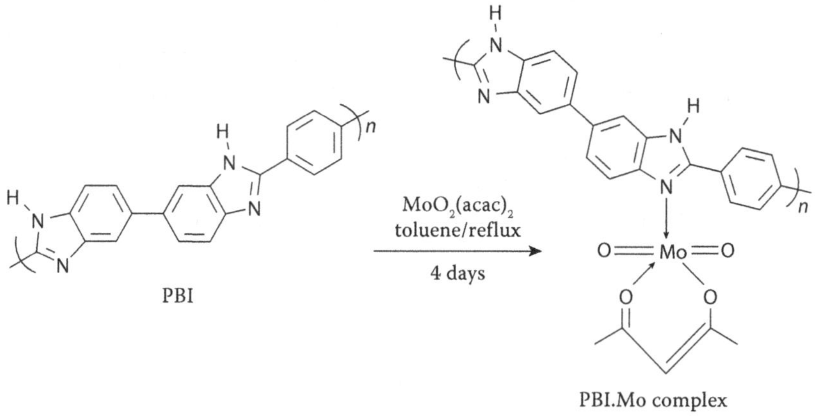

FIGURE 3.4 Synthesis of polybenzimidazole-supported Mo(VI) (PBI.Mo) complex.

washed with deionized water until the pH of the washings turned neutral, and then washed with acetone before being dried in vacuum at 313 K.

The polymer-supported Mo(VI) catalyst (PBI.Mo) was prepared by using a ligand exchange procedure in which the treated PBI resin was reacted with an excess of $MoO_2(acac)_2$ in the stoichiometric ratio of 2:1 $MoO_2(acac)_2$ to a functional ligand. For instance, 5 g of the treated PBI resin was refluxed with 17.68 g of $MoO_2(acac)_2$ in anhydrous toluene. The loading of Mo(VI) complex onto PBI resin was carried out using the same procedure as described in Section 3.2.1.3. The PBI beads changed from brown to green during this period. At the end of the reaction, the PBI.Mo catalyst particles were filtered off and extracted exhaustively with acetone to remove the excess $MoO_2(acac)_2$ in a similar way as described in Section 3.2.1.3. The catalyst was then dried in a vacuum oven at 313 K. The reaction is illustrated in Figure 3.4.

3.2.3 Catalysts Characterization

The polymer-supported Mo(VI) catalysts, that is, Ps.AMP.Mo and PBI.Mo, have been characterized extensively to determine their molecular structure, morphological, and physicochemical properties. The infrared spectra of the Ps.AMP.Mo and PBI.Mo catalysts were observed on a Thermal Nicolet Avectar 370 DTGS. The spectrum of the Ps.AMP.Mo catalyst (Figure 3.5) has revealed some bands at the range of ~760 to ~800 cm⁻¹, which are characteristics of Mo=O symmetrical and antisymmetric stretches. However, the band recorded at ~700 cm⁻¹ could most likely be associated with stretching Mo−O−Mo bridges.[52] Similarly, the PBI.Mo spectrum showed Mo=O and Mo−O−Mo vibration characteristics with adsorptions around ~690 to ~900 cm⁻¹ (Figure 3.5). Therefore, the FTIR spectra of both the Ps.AMP.Mo and PBI.Mo catalysts confirm the incorporation of Mo centers in both resins due to the presence of Mo=O and Mo−O−Mo features.

The molybdenum content of the prepared catalysts was analyzed using a Perkin–Elmer AAnalyst 200 spectrophotometer. A sample of each polymer-supported complex (~0.1 g) was grounded to a fine powder and digested in 15 mL aqua regia for 3 days. Aqua regia is a mixture of concentrated HNO_3 and HCl in the volume ratio of 1:3. The resulting mixture was diluted to 100 mL with distilled water and the Mo content was analyzed using an atomic absorption spectrophotometer (AAS). The Mo content of the Ps.AMP.Mo and PBI.Mo catalysts and the corresponding ligand:Mo ratio are given in Table 3.1.

The morphologies of the Ps.AMP.Mo and PBI.Mo catalyst particles were observed using a JEOL JSM-6300F scanning electron microscope (SEM). The SEM images of

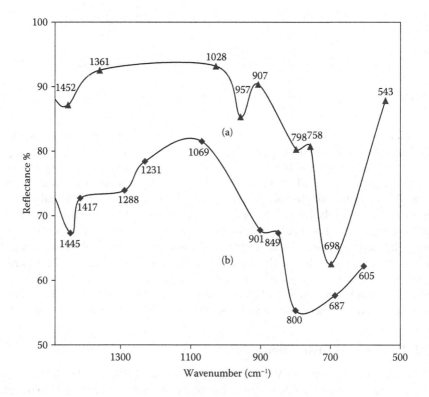

FIGURE 3.5 FTIR spectra of (a) Ps.AMP.Mo and (b) PBI.Mo catalysts.

TABLE 3.1

Physical and Chemical Properties of Ps.AMP.Mo and PBI.Mo Catalysts

Catalyst Properties	Ps.AMP.Mo Catalyst	PBI.Mo Catalyst
Average density (g/cm³)	1.44	1.74
BET surface area (m²/g)	53.5	24.8
Mo loading (mmol Mo/g resin)[a]	0.74	0.93
Ligand loading (mmol/g resin)[b]	0.5	2.4
Ligand to Mo ratio	0.68:1	2.58:1
Particle size (μm)	119–153	243–335
Total pore volume (cm³/g)	0.08	0.07
Average pore diameter (nm)	6.0	11.5

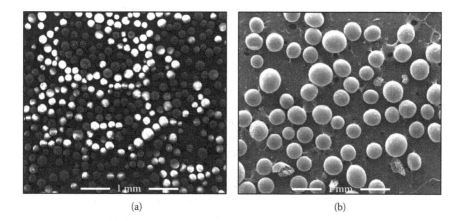

(a)   (b)

FIGURE 3.6 SEM images of (a) Ps.AMP.Mo and (b) PBI.Mo catalysts.

Ps.AMP.Mo and PBI.Mo (Figure 3.6) revealed an obvious similarity in the morphology of the two polymer supports. Neglecting mechanical damages to the samples, the SEM images of both catalysts have a well dispersed spherical smooth surface, a characteristic of macroporous polymer resins.[52] Apparently, the sizes of PBI.Mo catalyst beads (Figure 3.6b) are approximately double the size of those of the Ps.AMP. Mo catalyst (Figure 3.6a).

The particle size measurement was carried out using a Malvern Mastersizer, and the true densities of the catalysts were measured using micrometrics multivolume pycnometer-1305. As shown in Table 3.1, the Ps.AMP.Mo catalyst has particle sizes in the range of 119–153 μm, which are smaller than the PBI.Mo catalyst with a particle size range of 243–335 μm. The average densities of Ps.AMP.Mo and PBI.Mo catalyst particles were found to be 1.44 and 1.74 g/cm³, respectively. The Brunauer–Emmett–Teller (BET) surface area, pore volume, and pore diameter were determined by a nitrogen adsorption and desorption method using a Micromeritics ASAP (accelerated surface area and porosimetry) 2010. The results obtained for the BET surface area, pore volume, and pore diameters of the Ps.AMP.Mo and PBI.Mo catalysts are presented in Table 3.1.

3.3 BATCH EPOXIDATION STUDIES

Batch epoxidation experiments have been carried out in a classical batch reactor to evaluate the catalytic activities and stabilities of the Ps.AMP.Mo and PBI.Mo catalysts using TBHP as an oxidant. The suitability and efficiency of both catalysts for alkene epoxidation have been compared by studying the effect of various parameters such as reaction temperature, FMR of alkene to TBHP, and catalyst loading on the yield of epoxide for optimizing reaction conditions in a batch reactor. A detailed evaluation of molybdenum (Mo) leaching from the polymer-supported catalysts has been conducted by assessing the catalytic activity of the residue obtained from the supernatant solutions of the reaction mixture after removal of the heterogeneous catalyst. The results of batch epoxidation form the basis for continuous epoxidation experiments in a FlowSyn reactor.

3.3.1 PREPARATION OF TBHP SOLUTION

TBHP solution in water (70% w/w) was purchased from Sigma–Aldrich Co. Ltd, and the water content was removed by a Dean–Stark apparatus from a toluene solution following a modified procedure that was previously reported by Sharpless and Verhoeven.[51] For instance, 130 mL of TBHP solution in water (70% w/w) and 160 mL of toluene were placed in a 0.5 L separating funnel. The mixture was swirled vigorously for about 1 min and allowed to settle, forming organic and aqueous layers. Approximately 30 mL of the aqueous layer was removed from the mixture and the remaining portion (organic layer) was transferred into a 0.5 L doubled-necked round bottom flask equipped with a thermostat, a Dean–Stark apparatus, and a reflux condenser. The flask was immersed in paraffin oil contained in a 1 L Pyrex glass basin. The mixture was heated on a hot plate via the paraffin oil. A magnetic stirrer was placed both in the mixture and in the paraffin oil to enhance heat distribution. The solution was refluxed for about 2 h and the temperature of the paraffin oil was kept at 413–418 K using a thermostat. The distillate, which is mostly water, was collected from the side arm and the remaining solution of anhydrous TBHP in toluene was cooled at room temperature and stored over 4A molecular sieves in a fridge at 275–281 K. The concentration of TBHP in the toluene solution was determined by iodometric titration.[51]

3.3.2 BATCH EPOXIDATION PROCEDURE

Batch epoxidation of alkenes with TBHP as an oxidant in the presence of a polymer-supported Mo(VI) catalyst was conducted in a 0.25 L jacketed four neck glass reactor. The batch reactor was equipped with a condenser, overhead stirrer, digital thermocouple, sampling point, and water bath.

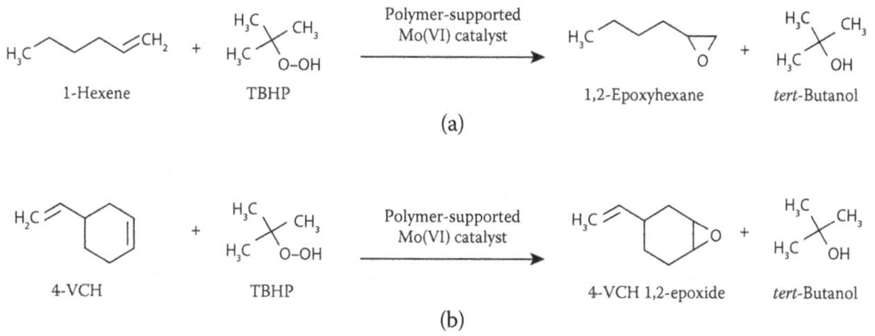

FIGURE 3.7 Reaction scheme for epoxidation of (a) 1-hexene and (b) 4-VCH with TBHP catalyzed by polymer-supported Mo(VI) catalyst.

Known quantities of alkene and TBHP were weighted out and introduced into the reactor vessel and stirring was started. The FMR of alkene to TBHP of 1:1–10:1 was selected for charging the reactor and agitation was started at the desired rate (400 rpm). Heating to the reaction mixture was supplied through water bath via the reactor jacket and monitored by digital thermocouple. The temperature of the reaction mixture was allowed to reach the desired value, that is, 323–343 K, and was maintained in the range of ±0.5 K throughout the batch experiment. A known amount of catalyst (0.15–0.6 mol% Mo loading) was added into the reactor when the reaction mixture reached a constant desired temperature. A sample was collected after the catalyst was added and the time was noted as zero time, that is, $t = 0$. Subsequent samples were taken from the reaction mixture at specific times and analyzed using Shimadzu GC-2014 gas chromatography (GC). The reactions for epoxidation of the model alkenes are illustrated in Figure 3.7.

3.3.3 METHOD OF ANALYSIS

Shimadzu GC-2014 GC was used for analyzing samples collected during epoxidation experiments. The instrument was fitted with a flame ionization detector, an auto-injector, and a 30 m long Econo-CapTM-5 (ECTM-5) capillary column (purchased from Alltech Associates, Inc., Deerfield, IL, USA) with internal diameter 320 μm and film thickness 0.25 μm. The carrier gas used was helium at a flow rate of 1 mL/min.

A split ratio of 100:1 and injection volume of 0.5 μL were selected as part of the GC method. Both injector and detector temperatures were maintained at 523 K. A ramp method was developed to separate all the components in the sample. In the ramp method, the oven temperature was initially set at 313 K, and the sample was then injected by the auto injector. The oven temperature was maintained at 313 K for 4 min after the sample was injected and ramped from 313 to 498 K at a rate of 293 K/min. Each sample took ~13 min to be analyzed by the GC and the oven temperature was cooled back to 313 K before the next run was started.

3.3.4 Batch Epoxidation Results

Batch epoxidation of 1-hexene and 4-VCH with TBHP as an oxidant has been carried out under different reaction conditions to study the effect of reaction temperature, FMR of alkene to TBHP, and catalyst loading on the yield of epoxide. Reusability and supernatant studies have been carried out to evaluate the long-term stability of each of the catalysts as well as the leaching of Mo from the polymer support. Both the Ps.AMP.Mo and PBI.Mo catalysts have been found to be selective in the formation of 1,2-epoxyhexane and 4-VCH 1,2-epoxide in the epoxidation of 1-hexene and 4-VCH with TBHP, respectively. The GC analysis of the reaction mixtures for 4-VCH epoxidation showed no evidence of either terminal or diepoxide products.

3.3.4.1 Investigation of Mass Transfer Resistances

Two types of mass transfer resistances exist in heterogeneous catalyzed alkene epoxidation with TBHP. One is across the solid–liquid interface, that is, the influence of the external mass transfer resistance caused as a result of stirring the reaction mixture. The other mass transfer resistance occurs in the intraparticle space, that is, internal mass transfer resistance that is connected with the different catalyst particle sizes and catalyst internal structures such as the chemical structure, pore size distribution, and porosity. A jacketed stirred batch reactor was used to study the existence of mass transfer resistance for alkene epoxidation with TBHP catalyzed by polymer-supported Mo(VI) catalysts. It was observed that there was negligible external mass transfer resistance when epoxidation experiments were carried out using a stirrer speed of 300–400 rpm under otherwise identical conditions. Therefore, it can be concluded that external mass transfer resistances were absent in this study. However, most of the particles of Ps.AMP.Mo and PBI.Mo lie within the size range of 119–335 μm, which are fairly uniform. According to Clerici and Kholdeeva,[53] mass transfer limitation could be eliminated when the catalyst particles are small and fairly uniform. Therefore, it was presumed that internal mass transfer resistance would be negligible for both Ps.AMP.Mo- and PBI.Mo-catalyzed epoxidation reactions due to the nature of the catalyst particles. On the basis of these investigations, all batch epoxidation experiments were carried out with a stirrer speed of 400 rpm using Ps.AMP.Mo and PBI.Mo catalysts as prepared.

3.3.4.2 Effect of Reaction Temperature

Alkene epoxidation with alkyl hydroperoxides essentially requires a thorough screening of reaction temperature in order to achieve high conversion of the oxidant and high product selectivity. Hence, epoxidation of 1-hexene and 4-VCH with TBHP has been carried out at 333, 343, and 353 K to study the effects of reaction temperature on the yield of corresponding epoxide.

As expected, higher reaction temperatures gave higher yields of epoxide at fixed reaction times for all the alkenes studied. Figure 3.8a shows that the yield of 1,2-epoxyhexane using the Ps.AMP.Mo catalyst at 260 min was 83% and 88% at 343 and 353 K, respectively, while a significant drop in the yield of

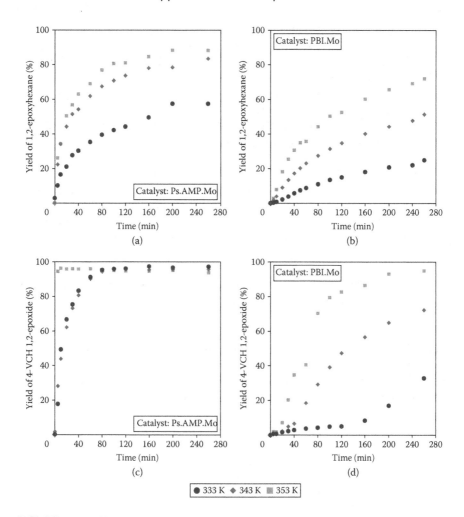

FIGURE 3.8 Effect of reaction temperature on the yield of epoxide for epoxidation of 1-hexene catalyzed by (a) Ps.AMP.Mo and (b) PBI.Mo catalysts, and 4-VCH catalyzed by (c) Ps.AMP.Mo and (d) PBI.Mo catalysts at catalyst loading: 0.3 mol% Mo, feed molar ratio of alkene to TBHP: 5:1, stirrer speed: 400 rpm.

1,2-epoxyhexane (57%) was recorded over the same period when the experiment was carried out at 333 K. Similar trends in the yield of 1,2-epoxyhexane (Figure 3.8b) and the yield of 4-VCH 1,2-epoxide (Figure 3.8d) were observed when the PBI.Mo catalyst was used. However, epoxidation of 4-VCH 1,2-epoxide in the presence of Ps.AMP.Mo reached equilibrium within the first 5 min for the experiment carried out at 353 K as shown in Figure 3.8c. This was due to a distinct exothermic effect observed during that period. The temperature was controlled immediately to maintain the reaction at 353 K. Therefore, the significantly higher yield of 4-VCH 1,2-epoxide obtained at that temperature was probably due to the exothermic effect. The yield of 4-VCH 1,2-epoxide during the first 5 min at 353 K was ~94%.

The effect was not so noticeable at 343 and 333 K. However, the yield of 4-VCH 1,2-epoxide obtained at 260 min was ~95% for all the three temperature ranges. The experiments were replicated twice and the same behavior was observed in both cases. Ambroziak et al.[5] observed a similar exothermic effect while studying the effect of reaction temperature on cyclohexene epoxidation with TBHP catalyzed by a polymer-supported Mo(VI) complex. Therefore, it can be concluded that 353 K is the preferred reaction temperature for both Ps.AMP.Mo- and PBI.Mo-catalyzed epoxidations of 1-hexene and 4-VCH (Figure 3.8a–d).

3.3.4.3   Effect of FMR

In most of the catalyzed alkene epoxidation processes, reactions are conducted with a substantially excess of alkene in order to avoid overoxidation and achieve high conversion of the oxidant and high yield of epoxide. Consequently, a number of batch experiments have been carried out to study the effect of different FMRs of alkene to TBHP on the yield of epoxide.

In the case of Ps.AMP.Mo-catalyzed epoxidation of 1-hexene and 4-VCH, an increase in the FMR of alkene to TBHP from 2.5:1 to 10:1 resulted in a marked increase in the yield of 1,2-epoxyhexane (Figure 3.9a) and 4-VCH 1,2-epoxide (Figure 3.9c). For instance, the yields of 1,2-epoxyhexane obtained at 260 min for 2.5:1 and 10:1 molar ratios of 1-hexane to TBHP were 61% and 97%, respectively (Figure 3.9a). The experiments conducted at 2.5:1 and 10:1 molar ratios of 4-VCH to TBHP gave 92% and 98% yields of 4-VCH 1,2-epoxide, respectively, at 260 min (Figure 3.9c). Similarly, a significant increase in the rate of epoxide formation was observed when the FMR of alkene to TBHP was increased from 2.5:1 to 5:1 for PBI.Mo-catalyzed epoxidation of 1-hexene (Figure 3.9b) and 4-VCH (Figure 3.9d). Surprisingly, for PBI.Mo-catalyzed epoxidation of 1-hexene (Figure 3.9b) and 4-VCH (Figure 3.9d), an increase in the FMR of alkene to TBHP from 5:1 to 10:1 decreases the rate of formation of corresponding epoxide, which is unusual. However, similar unexpected results were reported by Ambroziak et al.[5] for 1-octene epoxidation with TBHP in the presence of a polymer-supported Mo(VI) catalyst when the molar ratio of 1-octene to TBHP was increased from 5:1 to 10:1.

Based on the results obtained from this study, it can be concluded that the effect of alkene to TBHP feed ratio is dependent not only on the type of alkene but also on the catalyst used. In the case of PBI.Mo-catalyzed epoxidation of 1-hexene and 4-VCH, increasing the FMR of alkene to TBHP beyond 5:1 decreases the TBHP concentration to such an extent as to cause a reduction in the yield of epoxide. Therefore, the FMR of 10:1 (alkene to TBHP) can be considered as the optimum for Ps.AMP. Mo-catalyzed epoxidation of 1-hexene and 4-VCH (Figure 3.9a and c), while a FMR of 5:1 can be regarded as the optimum for the epoxidation of both alkenes in the presence of PBI.Mo catalyst (Figure 3.9b and d).

3.3.4.4   Effect of Catalyst Loading

In this chapter, the catalyst loading was defined based on the active Mo content instead of the total mass of the polymer-supported catalysts in order to take into account the slight differences in Mo loading that could arise from different batches

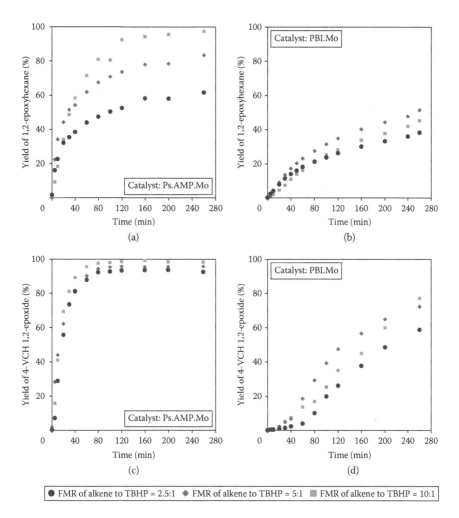

FIGURE 3.9 Effect of feed molar ratio of alkene to TBHP on the yield of epoxide for epoxidation of 1-hexene catalyzed by (a) Ps.AMP.Mo and (b) PBI.Mo catalysts, and 4-VCH catalyzed by (c) Ps.AMP.Mo and (d) PBI.Mo catalysts at reaction temperature: 343 K, catalyst loading: 0.3 mol% Mo, and stirrer speed: 400 rpm.

of the prepared catalysts. However, all the experiments in this study were conducted from one batch of the prepared catalysts.

An increase in catalyst loading increases the number of active sites per unit volume of reactor leading to an increase in the yield of epoxides. Thus, the effect of catalyst loading (i.e., mole ratio of Mo to TBHP × 100%) for epoxidation of 1-hexene and 4-VCH with TBHP was investigated by conducting batch experiments using 0.15 mol% Mo, 0.3 mol% Mo, and 0.6 mol% Mo catalyst loading. Epoxidation of 1-hexene catalyzed by the Ps.AMP.Mo catalyst shows identical trends in the rate of epoxidation when the catalyst loading was increased from 0.3 to 0.6 mol% Mo as shown in Figure 3.10a. However, 4-VCH epoxidation in the presence of Ps.AMP.Mo reached equilibrium within the first 20 min for catalyst loading of 0.6 mol% Mo, while it took 100 min to achieve equilibrium for the reaction conducted at 0.3 mol% Mo loading (Figure 3.10c). It should be noted that the Ps.APM.Mo catalyst demonstrates a remarkable catalytic performance for 4-VCH epoxidation as is evident by the higher yield of 4-VCH 1,2-epoxide (~90%) obtained at 260 min using 0.15 mol% Mo (Figure 3.10c) compared to the yield of epoxide achieved with other experiments carried out at 0.15 mol% Mo (Figure 3.10a, b, and d). It can be concluded from Figure 3.10a through d that the catalyst loading of 0.6 mol% Mo was found to be the optimum for both Ps.AMP.Mo- and PBI.Mo-catalyzed epoxidations of 1-hexene and 4-VCH.

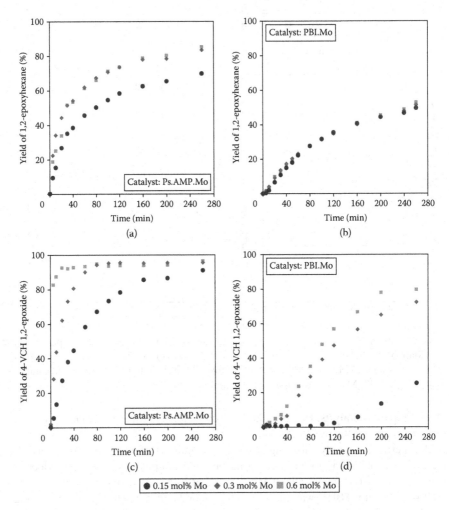

FIGURE 3.10 Effect of catalyst loading on the yield of epoxide for epoxidation of 1-hexene catalyzed by (a) Ps.AMP.Mo and (b) PBI.Mo catalysts, and 4-VCH catalyzed by (c) Ps.AMP. Mo and (d) PBI.Mo catalysts at reaction temperature: 343 K, feed molar ratio of alkene to TBHP: 5:1, and stirrer speed: 400 rpm.

3.3.4.5　Catalyst Reusability Studies

Transition metal catalysts that could be reused several times without significant loss in activity are generally attractive for commercial applications as most of the metal complexes are very expensive to purchase and difficult to prepare. The reusability potentials of the Ps.AMP.Mo and PBI.Mo catalysts for 1-hexene and 4-VCH epoxidation have been investigated by recycling the catalysts several times in batch experiments. The results are presented in Figure 3.11.

In this study, a fresh catalyst was used for epoxidation experiments and plotted as Run 1. At the end of the experiment, that is, Run 1, the catalyst particles were filtered

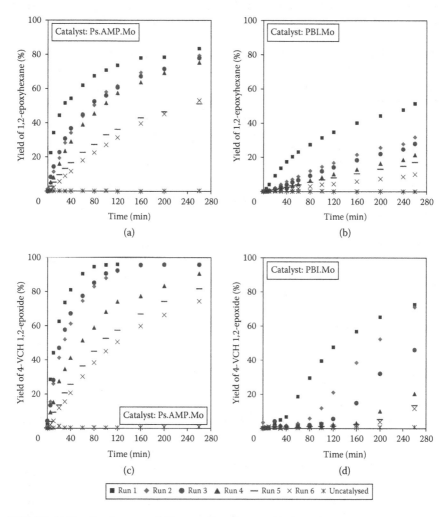

FIGURE 3.11 Catalyst reusability studies for epoxidation of 1-hexene catalyzed by (a) Ps.AMP.Mo and (b) PBI.Mo catalysts, and 4-VCH catalyzed by (c) Ps.AMP.Mo and (d) PBI.Mo catalysts at reaction temperature: 343 K, catalyst loading: 0.3 mol% Mo, feed molar ratio of alkene to TBHP: 5:1, and stirrer speed: 400 rpm.

from the reaction mixture, washed carefully with 1,2-dichloroethane, and stored in a vacuum oven at 313 K. The stored catalyst particles were reused in the subsequent experiment and plotted as Run 2. This procedure was repeated for the successive catalyst reusability experiments, that is, Run 3 to Run 6. In addition, an uncatalyzed epoxidation experiment was carried out and plotted for comparison with the catalyzed experimental results. As shown in Figure 3.11a through d, a high rate of epoxidation was observed in Run 1 as compared to subsequent runs. This is due to the sufficient active sites that are available for adsorption by the reacting species in the fresh catalyst sample compared with subsequent runs. The rate of formation of epoxide in Run 2 and Run 3 was quite similar in the Ps.AMP.Mo-catalyzed epoxidation of 1-hexene (Figure 3.11a) and 4-VCH (Figure 3.11c). The yield of 1,2-epoxyhexane obtained at 260 min was ~80% in both Run 2 and Run 3 (Figure 3.11a), while ~96% yield of 4-VCH 1,2-epoxide was obtained for Run 1 to Run 3 over the same period (Figure 3.11c). However, a significant drop in the rate of epoxide formation was observed in the subsequent Run 4 to Run 6 in Ps.AMP.Mo-catalyzed epoxidation of 1-hexene and 4-VCH (Figure 3.11a and c). Similarly, for epoxidation in the presence of the PBI.Mo catalyst, the yield of 1,2-epoxyhexane and 4-VCH 1,2-epoxide was decreased to ~11% after Run 6 as compared to Run 1 (Figure 3.11b and d).

The most distinguishing parameter that differentiates the Ps.AMP.Mo catalyst from the PBI.Mo catalyst is the ligand to Mo ratio. In case of the Ps.AMP.Mo catalyst, the ligand to Mo ratio is 0.68:1, while it is 2.58:1 for the PBI.Mo catalyst. It is obvious that Ps.AMP.Mo has an excess of Mo to ligand content as compared to PBI.Mo. Therefore, Mo is lost during reaction with the Ps.AMP.Mo catalyst since a significant proportion of Mo introduced at the outset cannot be coordinated by the polymer-immobilized ligand. In the case of the PBI.Mo catalyst, it seems most of the Mo introduced at the outset is coordinated by a polymer-immobilized ligand, and the Mo species that becomes mobile during catalysis is effectively recaptured by the presence of excess ligands on the support. On the contrary, Mo species containing microgel are lost from both the Ps.AMP.Mo and PBI.Mo catalysts as a result of mechanical agitation. It could be concluded that the reason for the decrease in catalytic activity after each successive experimental run for both catalysts was due to the loss of catalytically active Mo from the polymer supports, either as soluble leached complexes or as traces of Mo containing microgels released as a result of mechanical attrition of the beads or both.

3.3.4.6 Supernatant Studies

The aim of this analysis is to investigate the extent of Mo leaching from polymer-supported catalysts. Once each reusability study experiment was completed (see Section 3.3.4.5), the catalyst particles were filtered out and the reaction mixture was vacuum distilled to recover the residue from the reaction supernatant solutions. The isolated residue from the supernatant solution of fresh catalyst was used as a potential catalyst for epoxidation and plotted as Run 1 in Figure 3.12. The same procedure was repeated for all the subsequent reusability studies and the corresponding residue obtained was used as the catalyst for the supernatant studies. Furthermore, a control experiment was carried out in the absence of residue and plotted in Figure 3.12 for comparison with the experiments carried out in the presence of residue.

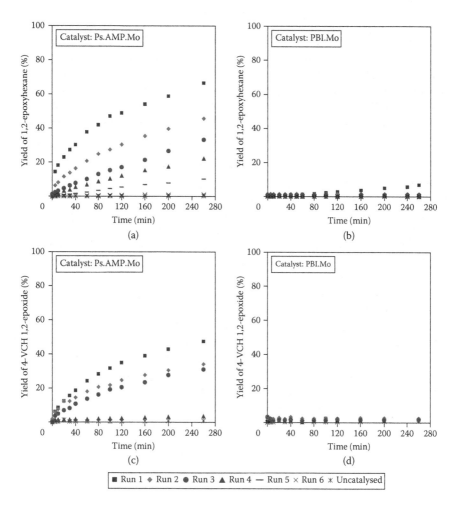

FIGURE 3.12 Supernatant studies for epoxidation of 1-hexene and 4-VCH catalyzed by residue isolated from supernatant solutions where polymer-supported Mo(VI) catalysts are reused at reaction temperature: 343 K, feed molar ratio of alkene to TBHP: 5:1, and stirrer speed: 400 rpm.

The residue isolated from Ps.AMP.Mo-catalyzed epoxidation of 1-hexene and 4-VCH revealed a high content of catalytically active Mo as shown in Figure 3.12a and c. This was evident by the catalytic effect observed when the experiments were conducted in the presence of residue. The amount of Mo retained by the catalyst after Run 1 was 0.70 mmol/g, while it was 0.62 and 0.56 mmol/g after Run 2 and Run 3, respectively, as compared to the Mo content in the fresh Ps.AMP.Mo, that is, 0.74 mmol/g resin (see Table 3.1). The yields of epoxide achieved from the supernatant studies of 1-hexene and 4-VCH after Run 1 were 67% (Figure 3.12a) and 47% (Figure 3.12c), respectively. The yields of epoxide decreased steadily for the subsequent experimental runs and became negligible after four to six runs (Figure 3.12a and c).

It can be seen from Figure 3.12b that the epoxidation of 1-hexene catalyzed by residue of PBI.Mo revealed the presence of some catalytically active Mo species after Run 1 as evident by the minor catalytic effect displayed by the reaction. However, the residue obtained from subsequent experimental runs show no signs of catalytic effect (Figure 3.12b). However, 4-VCH epoxidation in the presence of residue of PBI.Mo failed to show any trace of Mo leaching as shown in Figure 3.12d. Thus, it can be concluded that the leaching of Mo in Ps.AMP.Mo-catalyzed epoxidation becomes negligible after four to five experimental runs.

3.3.4.7 Optimization Study of Alkene Epoxidation

Further study aimed at achieving a set of optimum reaction conditions that would produce a maximum yield of epoxide for 1-hexene epoxidation have been carried out using the optimum reaction parameters. For 1-hexene epoxidation catalyzed by the Ps.AMP.Mo catalyst, the optimization study was conducted at a reaction temperature of 353 K, catalyst loading of 0.6 mol% Mo, and FMR of 1-hexene to TBHP of 10:1. In the case of PBI.Mo-catalyzed epoxidation of 1-hexene, the reaction temperature has the most pronounced effect on the yield of epoxide as shown in Figure 3.8b, while the influence of catalyst loading was not very distinct (Figure 3.10b). Furthermore, the reaction conducted at a FMR of 1-hexene to TBHP of 5:1 achieved a higher yield of epoxide than the one carried out at a FMR of 10:1 (Figure 3.9b). Hence, the optimization study for 1-hexene epoxidation in the presence of PBI.Mo was carried out at a reaction temperature of 353 K, catalyst loading of 0.6 mol% Mo, and FMR of 1-hexene to TBHP of 5:1. The maximum yield of 1,2-epoxyhexane obtained at the optimum reaction conditions for a Ps.AMP.Mo-catalyzed reaction at 260 min was ~97% (Figure 3.13a), while the yield of epoxide achieved at the same period with the optimum conditions using PBI.Mo was ~86% (Figure 3.13b).

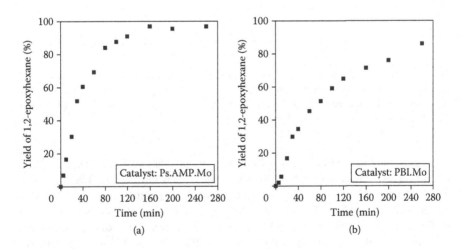

FIGURE 3.13 Optimization studies for 1-hexene epoxidation catalyzed by (a) Ps.AMP.Mo catalyst at a feed molar ratio of 1-hexene to TBHP: 10:1 and (b) PBI.Mo catalyst at feed molar ratio of 1-hexene to TBHP: 5:1, reaction temperature: 353 K, catalyst loading: 0.6 mol% Mo, and stirrer speed: 400 rpm.

3.4 CONTINUOUS ALKENE EPOXIDATION IN A FLOWSYN REACTOR

Continuous epoxidations of 1-hexene and 4-VCH have been carried out in a FlowSyn continuous flow reactor supplied by Uniqsis Ltd in the presence of Ps.AMP.Mo and PBI.Mo as catalysts. The instrument is equipped with two independent HPLC pumps, a control interface, and SquirrelView software with data logger (supplied by Grant Instruments). The catalytic fixed bed is made of a stainless steel (SS) column of length 130 mm (internal diameter 7 mm and outer diameter 10 mm). The fluid paths in the FlowSyn reactor were connected with each other using a perfluoropolymer tubing of 0.5 mm internal diameter. Each HPLC pump was primed before starting the experiment to remove air bubbles that may have been present in the fluid paths. All the reaction parameters including reaction temperature, feed flow rate, and pressure limits were set using the control interface of the FlowSyn reactor. The SquirrelView software and data logger were employed as additional components to the FlowSyn unit to precisely monitor and record the temperature profile of the mobile phase in the fixed-bed column. The SS column reactor was packed with the catalyst (1.5 ± 0.02 g) and enclosed in an electronically controlled column heater. The schematic of the experimental set-up of FlowSyn reactor is shown in Figure 3.14.

Continuous epoxidation studies were carried out following optimization of the reaction conditions as well as extensive evaluation of the activity and reusability of the heterogeneous catalysts for alkene epoxidation in a 0.25 L jacketed stirred batch reactor (see Section 3.3).

3.4.1 CONTINUOUS EPOXIDATION PROCEDURE IN A FLOWSYN REACTOR

Before starting the continuous epoxidation experiment using a FlowSyn reactor, the alkene was fed continuously by an HPLC pump until the column reactor and tubings were completely saturated. The heating to the column was set to the required value and the temperature of the mobile phase was allowed to reach the desired level and maintained at ±2 K. Once the column was saturated with alkene and the desired temperature was achieved, the continuous epoxidation experiment was started.

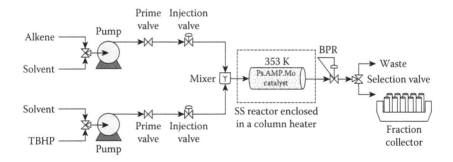

FIGURE 3.14 Schematic representation of continuous epoxidation of alkene using a FlowSyn reactor.

The reactants were continuously fed at a desired flow rate by two HPLC pumps to the packed column via a mixing chamber. As soon as the reactants entered the column, a reaction took place in the presence of the catalyst to produce epoxide. The fluid from the outlet port of the packed column was collected in a fraction collector. The samples were taken at specific time intervals and were analyzed by Shimadzu GC-2014 GC (see Section 3.3.3). The fluid paths in the FlowSyn reactor were properly cleaned by flushing with *iso*-propanol (solvent) at the end of each experimental run.

3.4.2 CONTINUOUS EPOXIDATION RESULTS

In a continuous epoxidation experiment using a FlowSyn reactor, a small volume of the reaction mixture is in contact with a relatively large volume of the catalyst in the SS column reactor. Moreover, particle size distribution is one of the most important properties that suggest the adaptability of a catalyst to a continuous flow reaction since fine powdered catalyst materials may cause high pressure drops in fixed bed reactors. The particle size range of the Ps.AMP.Mo catalyst (i.e., 119–153 μm) allows a stable system pressure during the reaction, thereby providing a constant flow of the feed through the reactor channels. Optimization of the reaction conditions have been carried out by studying the effects of different parameters such as reaction temperature, feed flow rate, and FMR of the alkene to TBHP on the conversion of TBHP and the yield of epoxide at steady state, that is, at 2 h. Figure 3.15 shows the steady-state mole fractions of the various constituents in the reaction mixture for continuous epoxidation of 4-VCH with TBHP as an oxidant. The long-term stability of the Ps.AMP.Mo catalyst for continuous epoxidation has been evaluated by reusing the same catalytic packing several times under similar reaction conditions. Furthermore, continuous epoxidations of 1-hexene and 4-VCH with TBHP have been carried out in

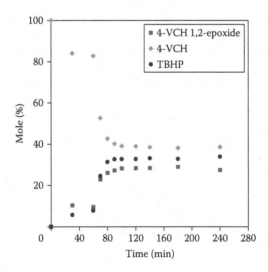

FIGURE 3.15 Mole fractions of the various constituents in the reaction mixture for continuous epoxidation of 4-VCH with TBHP as an oxidant.

the presence of PBI.Mo as a catalyst using the optimum reaction conditions recorded when Ps.AMP.Mo catalyst was used, so as to compare the catalytic performances of both catalysts for continuous epoxidation.

3.4.2.1 Effect of Reaction Temperature

Continuous epoxidations of 1-hexene and 4-VCH with TBHP in the presence of Ps.AMP.Mo as a catalyst were carried out at 333, 343, and 353 K to study the effect of reaction temperature on the conversion of TBHP and the yield of epoxide. The experiments were conducted using 5:1 molar ratio of alkene to TBHP and the feed flow rate was maintained at 0.1 mL/min. The temperature of the mobile phase in the catalytic column was monitored with the aid of SquirrelView software.

It can be seen from Figure 3.16a that ~45% conversion of TBHP and ~42% yield of 1,2-epoxyhexane was achieved at steady state for epoxidation of 1-hexene conducted at 333 K, while the reactions carried out at 353 K gave ~79% conversion of TBHP and ~64% yield of 1,2-epoxyhexane. The effect of reaction temperature on continuous epoxidation of 4-VCH with TBHP revealed a higher conversion of TBHP and yield of epoxide compared to 1-hexene epoxidation for all the temperature ranges studied as shown in Figure 3.16b. The conversion of TBHP and the yield of 4-VCH 1,2-epoxide at 333 K were found to be ~66% and ~64%, respectively (Figure 3.16b). However, a significant increase in the conversion of TBHP (~95%) and the yield of 4-VCH 1,2-epoxide (~82%) was obtained for reactions carried out at 353 K (Figure 3.16b). Hence, 353 K was selected for further optimization studies of 1-hexene and 4-VCH epoxidations in a FlowSyn reactor.

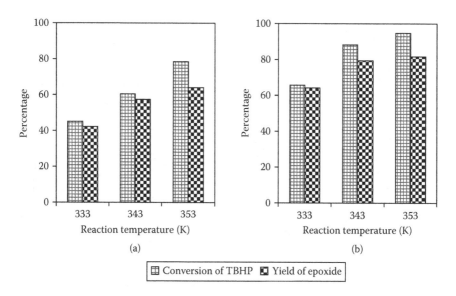

FIGURE 3.16 Effect of reaction temperature on the conversion of TBHP and the yield of epoxide at steady state for continuous epoxidation of (a) 1-hexene and (b) 4-VCH with TBHP using a FlowSyn continuous flow reactor in the presence of Ps.AMP.Mo catalyst (~1.5 g) at feed flow rate: 0.1 mL/min, feed molar ratio of alkene to TBHP: 5:1.

3.4.2.2 Effect of Feed Flow Rate

The effect of feed flow rate was investigated at 0.1, 0.13, and 0.16 mL/min. These flow rates correspond to the feed residence times in the reactor of ~5, ~4, and ~3 min, respectively. The experiments were carried out at 353 K (i.e., the optimum reaction temperature) and at a FMR of 5:1 (alkene to TBHP) using Ps.AMP.Mo as a catalyst.

An increase in feed residence time in the catalytic feed bed reactor by reducing the flow rate could have a positive impact on the catalytic performance in a continuous flow reaction. In the case of 1-hexene epoxidation, experiments carried out at a flow rate of 0.16 mL/min achieved ~68% conversion of TBHP and ~51% yield of 1,2-epoxyhexane as shown in Figure 3.17a. However, 4-VCH epoxidation at a feed flow rate of 0.16 mL/min gave ~84% conversion of TBHP and ~73% yield of 4-VCH 1,2-epoxide (Figure 3.17b). However, when the residence time of the feed was increased to ~4 min by reducing the flow rate to 0.13 mL/min, ~70% conversion of TBHP and ~55% yield of 1,2-epoxyhexane were recorded for 1-hexene epoxidation (Figure 3.17a), while the conversion of TBHP and the yield of 4-VCH 1,2-epoxide obtained at 0.13 mL/min increased to ~90% and ~77%, respectively, in the case of 4-VCH epoxidation (Figure 3.17b). Thus, it can be concluded that an increase in feed flow rate caused reduction in feed residence time in the reaction zone (packed column), which consequently led to a decrease in both the conversion of TBHP and the yield of corresponding epoxides in the continuous flow epoxidation.

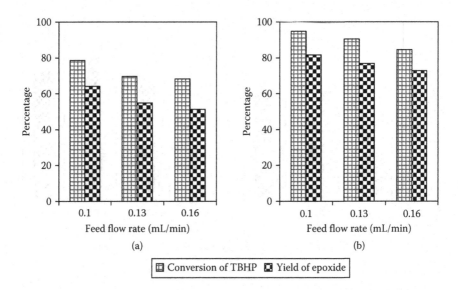

FIGURE 3.17 Effect of feed flow rate on the conversion of TBHP and the yield of epoxide at steady state for continuous epoxidation of (a) 1-hexene and (b) 4-VCH with TBHP using a FlowSyn continuous flow reactor in the presence of Ps.AMP.Mo catalyst (~1.5 g) at reaction temperature: 353 K, feed molar ratio of alkene to TBHP: 5:1.

3.4.2.3 Effect of FMR of Alkene to TBHP

The effect of FMRs of alkene to TBHP of 1:1, 2.5:1, and 5:1 were studied. The continuous epoxidation experiments were carried out at a feed flow rate of 0.1 mL/min and at a reaction temperature of 353 K. Figure 3.18 illustrates that both TBHP conversion and the yield of epoxide increase with an increase in FMR of alkene to TBHP. The experiment conducted at a FMR of 1-hexene to TBHP of 1:1 resulted in a similar TBHP conversion and yield of 1,2-epoxyhexane (~18%) as shown in Figure 3.18a. However, when a FMR of 1-hexene to TBHP of 2.5:1 was used, ~73% conversion of TBHP and ~63% yield of 1,2-epoxyhexane were achieved. Similarly, the reaction carried out at a FMR of 4-VCH to TBHP of 1:1 recorded ~50% conversion of TBHP and ~43% yield of 4-VCH 1,2-epoxide (Figure 3.18b). However, a significant increase in both the conversion of TBHP (~78%) and the yield of 4-VCH 1,2-epoxide (~70%) was obtained at a FMR of 4-VCH to TBHP of 2.5:1 (Figure 3.18b). Therefore, it can be concluded that the FMR of 5:1 (alkene to TBHP) is the appropriate molar ratio for continuous epoxidation of 1-hexene and 4-VCH with TBHP in the presence of Ps.AMP.Mo catalyst.

3.4.2.4 Catalyst Reusability Studies

Reusability studies of the Ps.AMP.Mo catalyst in batch epoxidations of 1-hexene and 4-VCH with TBHP were slightly affected by the attrition of catalyst particles and leaching of Mo containing microgel from the polymer support when the catalyst was reused under stirred conditions in a batch reactor (see Figure 3.11a and c). In continuous

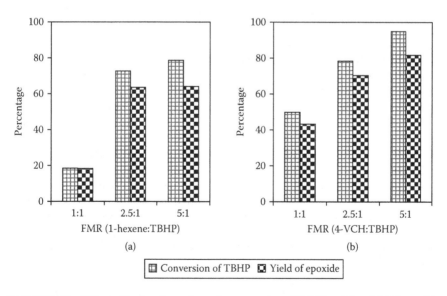

FIGURE 3.18 Effect of feed molar ratio on the conversion of TBHP and the yield of epoxide at steady state for continuous epoxidation of (a) 1-hexene and (b) 4-VCH with TBHP using a FlowSyn continuous flow reactor in the presence of Ps.AMP.Mo catalyst (~1.5 g) at reaction temperature: 353 K, feed flow rate: 0.1 mL/min.

epoxidation of 1-hexene and 4-VCH with TBHP using FlowSyn reactor, the Ps.AMP.Mo catalyst was firmly packed inside a SS column and there was no stirring involved. The reusability studies were carried out in a FlowSyn reactor using the same catalyst packing for four consecutive experimental runs, and each experiment lasted for 6 h. Therefore, the catalyst packing was used in continuous experiments for a period of 24 h. All the experiments for this study were carried out at a FMR of 0.1 mL/min, reaction temperature of 353 K, and FMR of alkene to TBHP of 5:1.

The reusability studies showed the Ps.AMP.Mo catalyst that was reused for four consecutive experimental runs under the same conditions had negligible loss in catalytic activity. In the case of 4-VCH epoxidations, the conversion of TBHP and the yield of 4-VCH 1,2-epoxide at steady state were found to be in the range of $95 \pm 4\%$ and $82 \pm 4\%$, respectively, for all the four experimental runs. However, 1-hexene epoxidation under similar conditions gave $79 \pm 4\%$ and $64 \pm 4\%$ conversions of TBHP and yields of 1,2-epoxyhexane, respectively, when the catalyst was reused four times. It could be concluded that the problems of attrition of catalyst particles and leaching of Mo observed in batch studies have been eliminated in continuous flow experiments. Therefore, the Ps.AMP.Mo catalyst could be reused several times for continuous epoxidations of alkene with TBHP in a fixed-bed column.

3.4.3 CONTINUOUS EPOXIDATION CATALYZED BY POLYBENZIMIDAZOLE-SUPPORTED MO(VI) COMPLEX (PBI.MO) IN A FLOWSYN REACTOR

It should be noted that continuous epoxidation of 1-hexene and 4-VCH catalyzed by Ps.AMP.Mo using a FlowSyn reactor achieved the highest conversion of TBHP and the yield of corresponding epoxides when the reactions were conducted at a reaction temperature of 353 K, feed flow rate of 0.1 mL/min, and FMR of alkene to TBHP of 5:1 (see Figure 3.17a and b). Hence, these conditions were employed for continuous epoxidation of 1-hexene and 4-VCH with TBHP as an oxidant using a FlowSyn reactor in the presence of PBI.Mo as a catalyst, so as to compare the performance of both polymer-supported catalysts for continuous epoxidation.

The results show that epoxidation of 1-hexene in the presence of PBI.Mo achieved ~35% conversion of TBHP and ~32% yield of 1,2-epoxyhexane, whereas the same reaction conducted in the presence of Ps.AMP.Mo as a catalyst gave ~79% conversion of TBHP and ~64% yield of 1,2-epoxyhexane (Figure 3.17a). However, continuous epoxidation of 4-VCH catalyzed by PBI.Mo gave appreciable conversion of TBHP (~84%) and yield of 4-VCH 1,2-epoxide (~69%), although the values are lower when compared with the conversion of TBHP (~95%) and the yield of 4-VCH 1,2-epoxide (~82%) obtained when Ps.AMP.Mo was used as a catalyst (Figure 3.17b). Therefore, it can be concluded that the Ps.AMP.Mo catalyst shows higher catalytic performance for continuous epoxidation of both 1-hexene and 4-VCH in a FlowSyn reactor as compared to PBI.Mo catalyst.

3.4.4 COMPARISON BETWEEN BATCH AND CONTINUOUS EPOXIDATION

In order to compare the efficiency of batch and continuous epoxidation, it is worth noting that epoxidation experiments were carried out in a batch reactor for a period of 240 min. However, a continuous epoxidation experiment in a FlowSyn reactor conducted at a feed flow rate of 0.1 mL/min has a feed residence time of ~5 min in the fixed-bed column. As shown in Figure 3.16a, continuous epoxidation of 1-hexene carried out in the presence of Ps.AMP.Mo at a reaction temperature of 343 K using a 5:1 molar ratio of 1-hexene to TBHP achieved ~58% yield of 1,2-epoxyhexane within ~5 min residence time, whereas it takes >40 min to achieve a similar yield of epoxide in a batch reactor using the same conditions (see Figure 3.8a). Similarly, ~80% yield of 4-VCH 1,2-epoxide was obtained for continuous epoxidation catalyzed by Ps.AMP.Mo at 343 K using a feed flow rate of 0.1 mL/min and a FMR of 4-VCH to TBHP of 5:1 (Figure 3.16b), while batch epoxidation of 4-VCH achieved similar yield of epoxide (80%) after 40 min of the reaction (Figure 3.8c). Therefore, based on the results obtained from this study, continuous epoxidation of alkene in a FlowSyn reactor can be considered to be more efficient than the batch reaction. Moreover, the continuous flow reaction in a FlowSyn reactor shows substantial benefits that include short setup and reaction times, flexibility of scaling-up reactions, and safer and more environmental friend operating procedures.

3.5   CONCLUSIONS

The catalytic performance of polymer-supported Mo(VI) complexes, that is, Ps.AMP.Mo and PBI.Mo, has been assessed for alkene epoxidation using TBHP as an oxidant. The Ps.AMP.Mo catalyst demonstrates better catalytic performance for epoxidation of 1-hexene and 4-VCH as compared to the PBI.Mo catalyst. For example, a Ps.AMP.Mo-catalyzed epoxidation of 4-VCH at 0.15 mol% Mo loading achieved ~91% yield of 4-VCH 1,2-epoxide at 260 min, whereas epoxidation of 4-VCH in the presence of PBI.Mo under similar conditions gave ~25% yield of epoxide over the same period. The catalyst reusability and supernatant studies have been carried out in a batch reactor to assess the long-term stability of the polymer-supported catalysts for alkene epoxidation using conditions that form the basis for continuous epoxidation experiments in a FlowSyn reactor. The supernatant studies confirmed the presence of some catalytically active Mo that might have contributed to catalysis in the Ps.AMP.Mo-catalyzed reactions. The leaching of Mo from the polymer support was due to soluble-leached complex or Mo-containing microgel released as a result of mechanical attrition of the beads or both.

Continuous alkene epoxidation using a FlowSyn reactor has enabled rapid evaluation of catalytic performance of Ps.AMP.Mo and PBI.Mo from a small quantity of reactants under different reaction conditions. Both catalysts have demonstrated higher catalytic performance in the continuous epoxidation of 4-VCH as compared to 1-hexene. However, this observation was not quite surprising due to the presence of an electron-withdrawing vinyl group in 4-VCH, which increased the reactivity of the double bond in the cyclic structure. Experiments carried out in the presence of Ps.AMP.Mo as a catalyst at a FMR of 4-VCH to TBHP of 5:1, reaction temperature of 353 K, and feed flow rate of 0.1 mL/min resulted in ~95% conversion of TBHP and ~82% yield of 4-VCH 1,2-epoxide at steady state. However, ~84% conversion of TBHP and ~69% yield of 4-VCH 1,2-epoxide were achieved for continuous epoxidation of 4-VCH catalyzed by PBI.Mo under similar reaction conditions. The results obtained in this study show that a thorough screening of reaction parameters including reaction temperature, feed flow rate, and FMR of alkene to TBHP could have a positive impact on the efficiency of a continuous flow alkene epoxidation in the presence of a heterogeneous catalyst.

ACKNOWLEDGMENTS

The financial support from EPSRC (grant nos EP/C530950/1 and EP/H027653/1) and The Royal Society Brian Mercer Feasibility award is gratefully acknowledged. M. L. Mohammed is grateful to Usmanu Danfodiyo University, Nigeria; Tertiary Education Trust Fund (TETFund), Nigeria; and London South Bank University, UK, for the PhD scholarships. The authors thank Dr. Rene Mbeleck and Dr. Krzysztof Ambroziak for their excellent contributions toward the success of the research work. We offer our sincerest gratitude to our former collaborator and an outstanding researcher, the late Professor David C. Sherrington.

REFERENCES

1. Sienel, G., Rieth, R., Rowbottom, K. T. Epoxides. In *Ullmann's Encyclopedia of Industrial Chemistry* (ed. Barbara Elvers). Wiley-VCH, Weinheim, 2000.
2. Bauer, K., Garbe, D., Surburg, H. *Common Fragrance and Flavour Materials*, Wiley-VCH, Weinheim, 2001, pp. 143–145.
3. Yudin, A. K. *Aziridines and Epoxides in Organic Synthesis*. Wiley-VCH, Weinheim, 2006, pp. 185–389.
4. Ambroziak, K., Mbeleck, R., Saha, B., Sherrinton, D. C. Epoxidation of limonene by *tert*-butyl hydroperoxide catalysed by polybenzimidazole-supported molybdcnum (VI) complex. *J. Ion Exchange* 2007, *18*, 598–603.
5. Ambroziak, K., Mbeleck, R., He, Y., Saha, B., Sherrington, D. C. Investigation of batch alkene epoxidations catalyzed by polymer-supported Mo(VI) complexes. *Ind. Eng. Chem. Res.* 2009, *48*, 3293–3302.
6. Ambroziak, K., Mbeleck, R., Saha, B., Sherrington, D. Greener and sustainable method for alkene epoxidations by polymer-supported Mo(VI) catalysts. *Int. J. Chem. React. Eng.* 2010, *8*, A12.
7. Mbeleck, R., Ambroziak, K., Saha, B., Sherrington, D. C. Stability and recycling of polymer-supported Mo(VI) alkene epoxidation catalyst. *React. Funct. Polym.* 2007, *67*, 1448–1457.
8. Mohammed, M. L., Mbeleck, R., Patel, D., Niyogi, D., Sherrington, D. C., Saha, B. Greener and efficient epoxidation of 4-vinyl-1-cyclohexene with polystyrene 2-(aminomethyl)pyridine supported Mo(VI) catalyst in batch and continuous reactors. *Chem. Eng. Res. Des.* 2014, *94*, 194–203. doi: 10.1016/j.cherd.2014.08.001.
9. Mohammed, M. L., Mbeleck, R., Patel, D., Sherrington, D. C., Saha, B. Greener route to 4-vinyl-1-cyclohexane 1,2-epoxide synthesis using batch and continuous reactors. *Green Process Synth.* 2014, *3*, 411–418.
10. Swern, D., Ed. *Organic Peroxides*, Wiley Interscience, New York, 1971.
11. Bezzo, F., Bertucco, A., Forlin, A., Barolo, M. Steady-state analysis of an industrial reactive distillation column. *Sep. Purif. Technol.* 1999, *16*, 251–260.
12. Liu, Y., Tsunoyama, H., Akita, T., Tsukuda, T. Efficient and selective epoxidation of styrene with TBHP catalyzed by Au-25 clusters on hydroxyapatite. *Chem. Commun.* 2010, *46*, 550–552.

13. Singh, B., Rana, B. S., Sivakumar, L. N., Bahuguna, G. M., Sinha, A. K. Efficient catalytic epoxidation of olefins with hierarchical mesoporous TS-1 using TBHP as an oxidant. *J. Porous Mat.* **2013**, *20*, 397–405.

14. Kollar, J. US Patent 3351635, 1967.

15. Arnold, U., Habicht, W., Doring, M. Metal-doped epoxy resins—New catalysts for the epoxidation of alkenes with high long-term activities. *Adv. Synth. Catal.* **2006**, *348*, 142–150.

16. Nath, G. R., Rajesh, K. Transition metal complexed crosslinked pyrazole functionalized resin-use as polymeric catalysts for epoxidation of olefin. *Asian J. Chem.* **2012**, *24*, 4548–4550.

17. Angelescu, E., Pavel, O. D., Ionescu, R., Birjega, R., Badea, M., Zavoianu, R. Transition metal coordination polymers MeX_2 (4,4′ bipyridine) (Me = Co, Ni, Cu, X = Cl, CH_3OCO-, acetylacetonate) selective catalysts for cyclohexene epoxidation with molecular oxygen and isobutyraldehyde. *J. Mol. Catal. A Chem.* **2012**, *352*, 21–30.

18. Gupta, K. C., Sutar, A. K. Catalytic activities of polymer-supported metal complexes in oxidation of phenol and epoxidation of cyclohexene. *Polym. Adv. Technol.* **2008**, *19*, 186–200.

19. Tangestaninejad, S., Mirkhani, V., Moghadam, M., Grivani, G. Readily prepared heterogeneous molybdenum-based catalysts as highly recoverable, reusable and active catalysts for alkene epoxidation. *Catal. Commun.* **2007**, *8*, 839–844.

20. Grivani, G., Akherati, A. Polymer-supported bis (2-hydroxyanyl) acetylacetonato molybdenyl Schiff base catalyst as effective, selective and highly reusable catalyst in epoxidation of alkenes. *Inorg. Chem. Commun.* **2013**, *28*, 90–93.

21. Mohammed, M. L., Patel, D., Mbeleck, R., Niyogi, D., Sherrington, D. C., Saha, B. Optimisation of alkene epoxidation catalysed by polymer-supported Mo(VI) complexes and application of artificial neural network for the prediction of catalytic performances. *Appl. Catal. A Gen.* **2013**, *466*, 142–152.

22. Mbeleck, R., Mohammed, M. L., Ambroziak, K., Sherrington, D. C., Saha, B. Efficient epoxidation of cyclododecene and dodecene catalysed by polybenzimidazole supported Mo(VI) complex. *Catal. Today* **2015**, *256*, 287–293.

23. Mohammed, M. L. *A Novel Heterogeneous Catalytic Process for the Synthesis of Commercially Important Epoxide Building Blocks.* PhD Thesis, London South Bank University, 2015.

24. Mohammed, M. L., Patel, D., Mbeleck, R., Sherrington, D. C., Saha, B. A safer and scalable continuous alkene epoxidation process. In *Proceedings of the 2nd International Congress of Chemical and Process Engineering, CHISA 2014*, Prague, Czech Republic, 23–27 August 2014.

25. Mohammed, M. L., Mbeleck, R., Patel, D., Sherrington, D. C., Saha, B. A greener, inherently safer and scalable continuous alkene epoxidation process. In *Proceedings of the ChemEngDayUK*, Imperial College London, UK, 25–26 March 2013.

26. Mohammed, M. L., Mbeleck, R., Patel, D., Sherrington, D. C., Saha, B. Greener and efficient alkene epoxidation process. In *Proceedings of the 9th European Congress of Chemical Engineering (ECCE9)*, The Hague, The Netherlands, 21–25 April 2013.

27. Mohammed, M. L., Mbeleck, R., Sherrington, D. C., Saha, B. Environmentally benign alkene epoxidation process. In *Proceedings of the 9th World Congress of Chemical Engineering (WCCE9)*, Coex, Seoul, Korea, 18–23 August 2013.

28. Saha, B. The Centre for Green Process Engineering (CGPE) opens with launch event at London South Bank University (LSBU). *Green Process Synth.* **2013**, *2*, 169–174.

29. Saha, B. Continuous clean alkene epoxidation process technology for the production of commercially important epoxide building bloc. In *Proceedings of the 3rd International Conference of the Flow Chemistry Society*, Munich, Germany, 19–20 March 2013.

30. Mbeleck, R., Mohammed, M. L., Ambroziak, K., Saha, B. Cleaner and efficient alkenes/terpenes epoxidation process catalysed by novel polymer supported Mo(VI) complexes. In *Proceedings of the Ion Exchange Conference (IEX 2012)*, Queens' College, University of Cambridge, UK, 19–21 September 2012.

31. Saha, B., Ambroziak, K., Sherrington, D. C., Mbeleck, R. US Patent 8759522 B2, 2014.

32. Dutia, P. Ethylene oxide: A techno-commercial profile. *Chem. Weekly* **2010**, pp. 199–203.

33. Dever, J. P., George, K. F., Hoffman, W. C., Soo, H. Ethylene oxide. In *Kirk-Othmer Encyclopedia of Chemical Technology* (ed. Arza Seidel), Wiley, New York, 2004, pp. 632–673.

34. Nijhuis, T. A., Makkee, M., Moulijn, J. A., Weckhuysen, B. M. The production of propene oxide: Catalytic processes and recent developments. *Ind. Eng. Chem. Res.* **2006**, *45*, 3447–3459.

35. Pham, H. Q., Marks, M. J. Epoxy resins. In *Ulmann's Encyclopedia of Industrial Chemistry* (ed. Barbara Elvers), Verlag Chemie, Weinheim, 2003.

36. Liao, Z. K., Boriack, C. J. US Patent 6087513, 2000.

37. Ansmann, A., Kawa, R., Neuss, M. *Cosmetic Composition Containing Hydroxyethers*, Cognis Deutschland, Düsseldorf, 2006.

38. Oyama, S. T. *Mechanisms in Homogeneous and Heterogeneous Epoxidation Catalysis*, Elsevier, Amsterdam, 2008.

39. Neri, G., Rizzo, G., Arico, A. S., Crisafulli, C., De Luca, L., Donato, A., Musolino, M. G., Pietropaolo, R. One-pot synthesis of naturanol from alpha-pinene oxide on bifunctional Pt-Sn/SiO$_2$ heterogeneous catalysts. Part I: The catalytic system. *Appl. Catal. A. Gen.* **2007**, *325*, 15–24.

40. Sell, C. *The Chemistry of Fragrance: From Perfumer to Consumer*, Royal Society of Chemistry, Dorset, 2006, pp. 70–71.

41. Wiemann, L. O., Faltl, C., Sieber, V. Lipase-mediated epoxidation of the cyclic monoterpene limonene to limonene oxide and limonene dioxide. *Z. Naturforsch. B.* **2012**, *67*, 1056–1060.

42. Liu, P., Jiang, L., Zhu, L., Wang, A. Novel covalently cross-linked attapulgite/poly (acrylic acid-*co*-acrylamide) hybrid hydrogels by inverse suspension polymerization: Synthesis optimization and evaluation as adsorbents for toxic heavy metal. *Ind. Eng. Chem. Res.* **2014**, *53*, 4277–4285.

43. Sherrington, D. C. Preparation, structure and morphology of polymer supports. *Chem. Commun.* **1998**, *30*, 2275–2286.

44. Holdge, P., Sherrington, D. C., Eds. *Polymer-Supported Reactions in Organic Synthesis*, Wiley, Chichester, 1980.

45. Mark, H. F., Bikales, N. M., Overberger, C. G., Menges, G., Kroschivitz, J. I., Eds. *Encyclopedia of Polymer Science and Engineering*, Wiley, New York, 1989.

46. Ward, B. C. Molded celazole PBI resin—Performance properties and aerospace-related applications. *SAMPE J.* **1989**, *25*, 21–25.

47. Yang, T., Shi, G. M., Chung, T. Symmetric and asymmetric zeolitic imidazolate frameworks (ZIFs)/polybenzimidazole (PBI) nanocomposite membranes for hydrogen purification at high temperatures. *Adv. Energy Mater.* **2012**, *2*, 1358–1367.

48. Grigoriev, S. A., Kalinnikov, A. A., Kuleshov, N. V., Millet, P. Numerical optimization of bipolar plates and gas diffusion electrodes for PBI-based PEM fuel cell. *Int. J. Hydrogen Energ.* **2013**, *38*, 8557–8567.

49. Ferng, Y. M., Su, A., Hou, J. Parametric investigation to enhance the performance of a PBI-based high-temperature PEMFC. *Energ. Convers. Manage.* **2014**, *78*, 431–437.

50. Zhu, W., Sun, S., Gao, J., Fu, F., Chung, T. Dual-layer polybenzimidazole/ polyethersulfone (PBI/PES) nanofiltration (NF) hollow fiber membranes for heavy metals removal from wastewater. *J. Membr. Sci.* **2014**, *456*, 117–127.

51. Sharpless, K. B., Verhoeven, T. R. Metal-catalyzed, highly selective oxygenations of olefins and acetylenes with *tert*-butyl hydroperoxide. Practical considerations and mechanisms. *Aldrichim. Acta* **1979**, *12*, 63–82.

52. Leinonen, S., Sherrington, D. C., Sneddon, A., McLoughlin, D., Corker, J., Canevali, C., Morazzoni, F., Reedijk, J., Spratt, S. B. D. Molecular structural and morphological characterization of polymer-supported Mo(VI) alkene epoxidation catalysts. *J Catal.* **1999**, *183*, 251–266.

53. Clerici, M. G., Kholdeeva, O. A. *Liquid Phase Oxidation via Heterogeneous Catalysis: Organic Synthesis and Industrial Applications.* Wiley, Hoboken, NJ, 2013.

4 Solid-Phase Heavy-Metal Separation with Selective Ion Exchangers
Two Novel Morphologies

Sukalyan Sengupta and Tabish Nawaz

CONTENTS

It is typical to encounter sludges and soils that are contaminated with heavy metals where the latter constitutes a small fraction (usually <5%) of the solid phase, with the rest being a background of innocuous nontoxic materials that are not important from a regulatory viewpoint. However, the regulatory protocol classifies the whole mass of soil/sludge as hazardous, resulting in extensive and expensive treatment to decontaminate this waste. If the target heavy metals can be selectively removed from the background matrix, this would constitute an efficient treatment process. In this context, though the background materials are nontoxic, they may interact with the heavy metals through generation of high buffer capacity, ion exchange, complexation, and so on, thus compounding the problem. This chapter explores the feasibility of using two novel morphologies of ion exchangers, composite ion exchange membranes and ion exchange fibers that can be used in such unfavorable conditions of solid-phase decontamination. The conventional morphology of ion exchangers, namely, beads and membranes, cannot be used with wastes that have high suspended solids. This chapter also reports on aspects of tailoring the chemistry within the sludge reactor to get optimum results. Results of detailed characterizations and extensive experiments using these two materials encompassing a wide range of possible scenarios are provided in this chapter.

4.1   INTRODUCTION

Various industrial processes produce wastes containing heavy-metal compounds along with high suspended solids (up to 10% mass/volume). The situation is similar with heavy-metal contaminated hazardous waste sites. In both cases, the heavy metals are most likely to be present in small amounts (<5% of the total mass of the solid phase) as precipitates of salts with low solubility products (e.g., as hydroxides, carbonates, sulfides), in the background of bulk amounts of innocuous solids such as sand, clay, calcite, and humin. Even though the percentage of heavy metals in the solid phase is very low, the solid waste is typically characterized as hazardous because the regulatory approach is based on almost total extraction of the metals in a limited volume of liquid and comparing this concentration against a defined limit. The concentration of metals is determined via total elemental analysis (USEPA Method 3050),[1] or by leach tests, such as the toxicity characteristic leaching procedure (TCLP) (USEPA Method 1311)[1] or by the synthetic precipitation-leaching procedure, or SPLP test (USEPA Method 1312).[1] In USEPA Method 3050, the solid waste is digested with a strong mineral acid and hydrogen peroxide and the digestate is analyzed for metal content, expressed as mg metal/kg^{-1} of the original solid. Leach test procedures measure the concentration of metals in leachate from soil contacted with an acetic acid solution (TCLP) or a dilute solution of sulfuric and nitric acid (SPLP). In this case, metal contamination is expressed in mg/L of the leachable metal. The regulatory limits are based on the premise that heavy-metal contamination of soil may pose risks and hazards to humans and the ecosystem through direct ingestion or contact with contaminated soil, the food chain (soil–plant–human or soil–plant-animal–human), drinking of contaminated ground water, reduction in food quality (safety and marketability) via phytotoxicity, reduction in land usability for agricultural production causing food insecurity, and land tenure problems.[2–5]

Naturally, these limits are strict; for example, the regulatory limit for cadmium in a solid waste is 85 mg/kg of the solid waste.[6] Thus, if one sample of the solid leaches cadmium at >85 mg/kg, the whole mass of the solid waste is considered hazardous. Naturally, this situation poses a major environmental challenge. Decontamination of heavy metals from this background can become economically feasible only if the heavy-metal ion is *selectively* removed from the rest of the solid phase that is composed of benign, nontoxic compounds. While the nontoxic materials are unimportant from a regulatory viewpoint, their chemical properties strongly influence the selective separation of heavy metals. From a broad environmental chemistry perspective, three cases are possible:

1. *Heavy-metal precipitates present with a chemically inert solid phase.* This case is the easiest in terms of removal of the heavy-metal ion. For example, if a $MeCO_3$ (s) is present in a background of *non-reacting* soil, addition of a mineral acid or an acidic gas such as CO_2 can strip the Me^{2+} from the matrix, thus increasing its aqueous-phase concentration. Scenario 1 presents the case of $PbCO_3(s)$ present in an inert background. Appendix 4.A lists all the simultaneous equations that need to be solved, and Figure 4.1a shows the pPb^{2+} profile as a function of partial pressure of $CO_2(g)$ applied. Figure 4.1b proves that the slope of pH versus $pCO_2(g)$ plot is 2/3 while that of pPb^{2+} versus $pCO_2(g)$ is 1/3. Both the figures are developed based on calculations presented in Appendix 4.A, and MINEQL+ 4.6[7] chemical equilibrium modeling software was used for the two plots. A subsequent solid–liquid separation will enable the remaining solid phase to be heavy-metal free and thus be classified as nonhazardous. The heavy-metal-rich aqueous solution can be treated using a variety of methods.

   Scenario 1: $PbCO_3(s)$ is the governing solid phase in an inert background and $CO_2(g)$ is introduced at a known partial pressure.

2. *Heavy metals present amidst a background or high buffer capacity.* In this case, the background solid-phase compounds can interact chemically with the heavy-metal dissolution. For example, if a heavy-metal precipitate such as $MeCO_3$ is present as a minor contaminant in a background of calcite soil, the total concentration of heavy metal in the aqueous phase, $[Me_t]$, will be orders of magnitude lower than that of the benign competing cation, $[Ca_t]$ in this case. Some conventional methods of increasing the aqueous-phase concentration of the heavy metal, for example, by adding a mineral acid or CO_2, may not provide any advantage since the competing ion will also rise in concentration, as shown in Scenarios 2 and 3.

   Scenario 2: $PbCO_3$ and $CaCO_3$ are the governing solid phases in a solid-phase background consisting of other benign, non-reacting compounds. $CO_2(g)$ is introduced into this system at a specified partial pressure. Appendix 4.B lists all the simultaneous equations that need to be solved, and Figure 4.2a shows the pPb^{2+}, pCa^{2+}, and pH profiles as a function of partial pressure of $CO_2(g)$ applied. Figure 4.2b proves that the slope of pH versus $pCO_2(g)$ plot is 2/3 while that of pPb^{2+} and pCa^{2+} versus $pCO_2(g)$ is 1/3. Both the figures are developed based on calculations presented in Appendix 4.B.

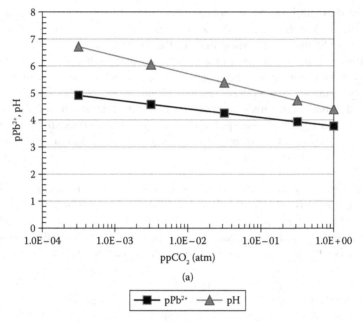

(a)

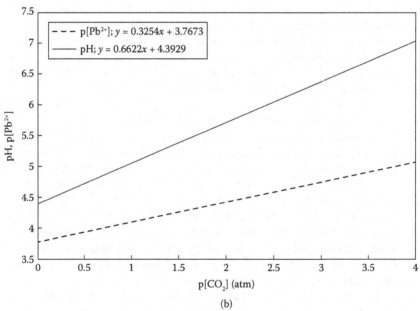

(b)

FIGURE 4.1   (a) Pb^{2+} concentration as a function of partial pressure of CO_2 for a sludge where $PbCO_3$ is the governing solid phase. (b) Slope of pPb^{2+} and pH versus $pCO_2(g)$ lines for case shown in Figure 4.1(a).

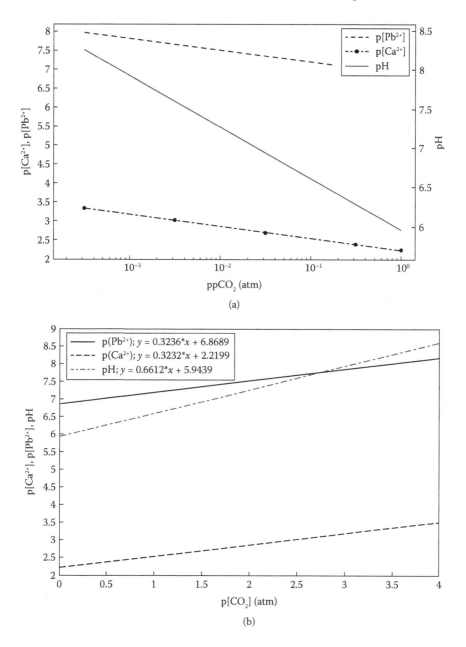

FIGURE 4.2 (a) Pb^{2+} and Ca^{2+} concentration as a function of partial pressure of CO_2 for a sludge where $PbCO_3$ and $CaCO_3$ are the governing solid phases. MINEQL+ 4.6[7] chemical equilibrium modeling software was used to generate this plot. (b) Slope of pPb^{2+}, pCa^{2+}, and pH versus $pCO_2(g)$ lines for case shown in Figure 4.2(a).

Scenario 3: PbCrO$_4$ and CaCO$_3$ are the governing solid phases in a solid-phase background consisting of other benign, non-reacting compounds. The difference from Scenario 2 is that both Pb and Cr are toxic heavy metals and need to be removed from the sludge. CO$_2$(g) is introduced into this system at a specified partial pressure. Figure 4.3, developed from calculations presented in Appendix 4.C, shows the profile of concentration of various pertinent species as a function of partial pressure of CO$_2$(g).

3. *Heavy-metal cations bound to the ion exchange sites of soil.* If the heavy-metal cation can associate with functional sites of the background material via adsorption or ion exchange, its removal entails a two-step process:
   a. Desorption of the heavy-metal cation from adsorption or ion exchange sites
   b. Selective removal of the heavy-metal cation from a background of innocuous cations, for example, Na$^+$, Ca^{2+}, and Mg^{2+}

Selective and targeted removal of the heavy metals from the background solid phase would constitute an efficient treatment process as it would render the sludge nonhazardous and also may make it possible for the heavy metals to be concentrated and recycled/reused. It is well known that chelating ion exchangers have high selectivity for heavy metals over alkali and alkaline earth metal cations, the competing species in such cases. However, the physical configuration of conventional ion exchangers (spherical or granular) makes their use inappropriate for such a case. We have identified two morphologies that show the potential of overcoming the challenge of selective heavy-metal cation separation from the background of a solid phase. The first material,[8–13] a new class of sorptive/desorptive composite ion exchange material (CIM), is

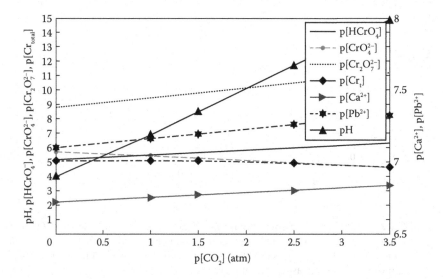

FIGURE 4.3 Pb^{2+}, Ca^{2+}, and various Cr species' concentration as a function of partial pressure of CO$_2$ for a sludge where PbCrO$_4$ and CaCO$_3$ are the governing solid phases. MINEQL+ 4.6[7] chemical equilibrium modeling software was used to generate this plot.

available commercially as thin sheets ($\approx$0.5 mm thick) and is suitable for heavy-metal decontamination from sludges and slurries. The morphology of the material—along with its physical texture and tensile strength—makes it compatible for use with sludges and slurries. The second material,[14] alkaline-hydrolyzed electrospun polyacrylonitrile (PAN) submicron fiber, termed ion exchange fiber (IXF), opens up a new area of ion exchanger synthesis and application, and can also be used to decontaminate heavy metal–laden sludges and slurries. This chapter presents details of material characterization of and potential application scenarios for both CIM and IXF.

4.2 COMPOSITE ION EXCHANGE MEMBRANE

4.2.1 CIM Characterization

The CIM is a thin sheet prepared by comminuting a cross-linked polymeric ion exchanger to a fine powder, and fabricating mechanically into a microporous composite sheet consisting of ion exchange powder matter enmeshed in polytetrafluoroethylene (PTFE). During this mechanical process, the PTFE microspheres are converted into microfibers that separate and enmesh the particles.[15] When dry, these composite sheets consist of >80% particles (polymeric ion exchanger) and <20% PTFE by weight. They are porous (usually >40% voids) with pore size distributions that are uniformly below 0.5 µm. The ion exchange microspheres are usually <100 µm in diameter and have a total thickness $\approx$0.5 mm. As such, they are effective filters that remove suspensoids >0.5 µm from permeating fluids. Because of such a sheet-like configuration, this material can be introduced easily into or withdrawn from a reactor with a high concentration of suspended solids, with the target solutes being adsorbed onto or desorbed from the microadsorbents. In this chapter, the chelating functionality of the microspheres chosen is iminodiacetate (IDA). Figure 4.4a shows an electron microphotograph of the composite IDA membrane, and Figure 4.4b provides a schematic depicting how the microbeads are trapped within the fibrous network of PTFE. Table 4.1 provides the salient properties of the composite IDA membrane used in the study. Note that the chelating microbeads constitute 90% of the composite membrane by mass. This feature allows the membrane to achieve the same level of performance as the parent chelating beads used in a fixed bed operation. More details about characterization of the membrane are

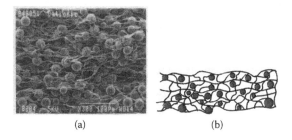

(a)   (b)

FIGURE 4.4 (a) Scanning electron microphotograph (300×) of the CIM. (b) Schematic representation of ion exchanger beads present in a network of microfibrous PTFE.

available in open literature[8,9,13] and are not repeated here. It may be noted that this material differs fundamentally from traditional ion exchange membranes used in industrial processes such as Donnan dialysis (DD) and electro dialysis (ED) because of its high porosity. DD and ED membranes have very low porosity and are strongly influenced by the Donnan Co-ion Exclusion Principle,[16] which does not allow anions to pass through a cation exchange membrane, and vice versa. However, in the case of a composite membrane, large gaps between ion exchangers allow anions to pass through freely even though it is a cation exchange membrane. The suspended solids that are >0.5 μm are not able to penetrate across the skin of the membrane because of the pore size of the material, as explained previously. However, water molecules and ions can easily move in and out of the thickness of the sheet, thus allowing for unimpeded ion exchange reactions between target ions in solutions (heavy metals in this case) and the counter ions of the membrane, as shown schematically in Figure 4.5 and explained in detail by Sengupta and SenGupta.[8,10,11,13] After a design time interval, the membrane can be taken out and chemically regenerated with a strong (3%–5%) mineral acid solution.

TABLE 4.1
Properties of the Composite Chelex Membrane

Composition	90% Chelex Chelating Resin, 10% Teflon
Pore size (nominal)	0.4 μm
Nominal capacity	3.2 meq/g dry membrane
Membrane thickness	0.4–0.6 mm
Ionic form (as supplied)	Sodium
Resin matrix	Styrene–divinylbenzene
Functional group	Iminodiacetate
pH Stability	1–14
Temperature operating range	0–75°C
Chemical stability	Methanol; 1 N NaOH, 1 N H_2SO_4
Commercial availability	Bio-Rad, Inc., Hercules, California

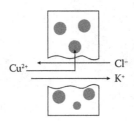

FIGURE 4.5   Schematic diagram showing the porosity of the CIM to cations and anions and selective entrapment of the heavy-metal cation.

4.2.2 PROCESS CONFIGURATION

Figure 4.6 shows a conceptualized process schematic where a composite membrane strip is continuously run through the contaminated sludge (sorption step) and an acid bath (desorption step). Such a cyclic process configuration is relatively simple and can be implemented by using the composite membrane as a slow-moving belt. For a sludge containing heavy-metal hydroxide, say $Me(OH)_2$, the process works in two steps:

1. *Sorption:* The CIM, when in contact with the sludge, selectively removes dissolved heavy metals from the aqueous phase in preference to other nontoxic alkali and alkaline earth metal cations. Consequently, fresh heavy-metal hydroxide dissolves to maintain equilibrium, and the following reactions occur in series:

$$\text{Dissolution: } Me(OH)_2(s) \leftrightarrow Me^{2+} + 2OH^- \qquad (4.1)$$

$$\text{CIM uptake: } \overline{RN(CH_2COOH)_2} + Me^{2+} \leftrightarrow \overline{RN(CH_2COO^-)_2 Me^{2+}} + 2H^+ \quad (4.2)$$

2. *Desorption:* When the CIM is immersed in the acid chamber, the exchanger microbeads are efficiently regenerated according to the following reaction:

$$\overline{RN(CH_2COO^-)_2 Me^{2+}} + 2H^+ \leftrightarrow \overline{RN(CH_2COOH)_2} + Me^{2+} \qquad (4.3)$$

The regenerated CIM is then ready for sorption again and the cycle is repeated. For such an arrangement to be practically feasible, the CIM sheet needs to be physically tough enough to withstand the tension of the conveyor belt and resilient to the chemical forces that are created during sorption and regeneration. Previous studies

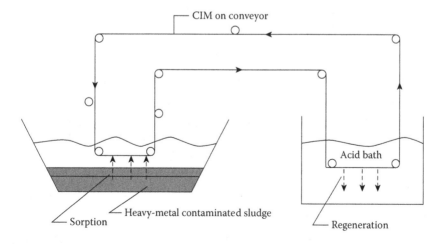

FIGURE 4.6 A conceptualized continuous decontamination process for heavy-metal removal from a sludge reactor using CIM.

conducted[17] in this area have confirmed that the CIM sheet is durable enough to withstand cyclical forces (physical and chemical) for higher than 200 cycles.

4.2.3 Cyclic Process for Heavy-Metal Extraction from Ion Exchange Sites of Soil

4.2.3.1 Experimental Procedure

To determine the viability of the CIM to extract heavy metals bound to the cation exchanging sites of soil, analytical grade bentonite was chosen as the soil type because of its high ion exchange capacity (IEC) and was loaded with Cu(II) by equilibrating it with an aqueous phase containing 400 mg/L Cu(II) concentration added as $Cu(NO_3)_2 \cdot 2.5H_2O$ at a pH of 5.5. The total exchange capacity of bentonite was found by mass balance. Cu(II)-loaded bentonite was then introduced into a plastic container containing 200 ml of 500 mg/L Na^+ (added as NaCl) solution. Cu(II)-loaded bentonite was the only solid phase present in the sludge and the suspended solids loading was 2.5% (m/v). A strip of conditioned CIM weighing 0.112 g was introduced into the plastic container and run for 30 cycles. In each cycle, the sorption step involved vigorously shaking the plastic container for 3 h at a constant pH of 5.0 while the desorption step used 5% H_2SO_4 (v/v) for 1 h for regeneration. The Cu(II) concentration in the regenerant was used to compute the percentage Cu(II) recovery from the bentonite phase.

4.2.3.2 Ion Selectivity of Bentonite

Since bentonite was chosen as the representative clay sample, ion selectivities of bentonite for Cu^{2+}, Ca^{2+}, and Na^+ were determined. Binary separation factors Cu/Na and Ca/Na were obtained experimentally by the following process:

For the Cu^{2+}/Ca^{2+} system:

1. Dry analytical grade bentonite was rinsed two to three times with deionized (DI) water to purge the bentonite slurry of any cations that may have been generated from the congruent/incongruent dissolution of any solid phase present in the original powder.
2. Next, 12 g of conditioned and dried bentonite was loaded in Ca^{2+} form by equilibrating it with a 600 mg/L Ca^{2+} (added as $Ca(NO_3)_2 \cdot 4H_2O$) at a pH ≈ 5.5. After an equilibrium period of 48 h, filtered solutions of initial and final slurry were analyzed for concentrations of Ca^{2+}, Mg^{2+}, and Al^{3+}. Calcium loading on the bentonite was determined by mass balance.
3. The calcium-loaded bentonite slurry was dewatered and dried. Different weights of the above-dried sample (1.75–2.5 g) were then equilibrated with 200 ml of 160 mg/L Cu^{2+} solution (present as $Cu(NO_3)_2 \cdot 2.5H_2O$) in plastic bottles where the pH in each bottle was ≈4.7.

4.2.3.3 Extraction Efficiency

Figure 4.7 shows the plot of percentage recovery of Cu(II) and the aqueous-phase Cu(II) concentration versus number of cycles for the case of Cu(II)-loaded bentonite. Note that >60% recovery of Cu(II) was achieved in less than 30 cycles.

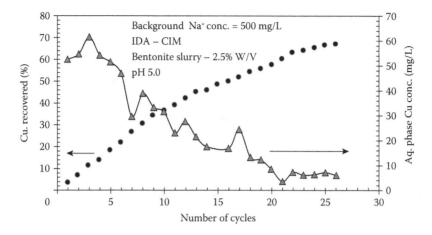

FIGURE 4.7   Copper (II) recovery from the ion exchange sites of bentonite clay during the cyclic process.

Bentonite clays (alumino silicates) derive their cation exchange capacity primarily from the isomorphous substitution within the lattice of aluminum and possibly phosphorus for silicon in tetrahedral coordination and/or magnesium, zinc, lithium, and so on, for aluminum in the octahedral sheet.[18] Divalent metal ions are strongly held onto these ion exchange sites primarily through electrostatic interactions. Removal of heavy metals from such contaminated soils during the sorption step essentially involves the following consecutive steps:

- Desorption from the ion exchange sites of the soil into the aqueous phase by a counter ion
- Selective sorption from the liquid phase onto the composite membrane

As in any sequential process with varying kinetic rates, the slower step will govern the rate of the overall process. Figure 4.8 compares the results of fractional desorption of Cu(II) from the ion exchange sites of bentonite at a pH of 5.0 using a 400 mg/L Ca(II) solution with that of fractional uptake of Cu(II) from a pure solution with an initial Cu(II) concentration of 200 mg/L carried out at the same pH value. From this figure, it can safely be concluded that the desorption rate of heavy metal from the ion exchange sites of clay is much faster than the sorption rate of the same by the composite membrane. For example, for the experimental conditions of Figure 4.8, the time taken for 50% desorption from bentonite is 6 min while that for 50% sorption by the membrane (both expressed as a ratio of the equilibrium capacity) is $\approx$240 min. Thus,

$$\frac{t_{\frac{1}{2}\ \text{sorption}}}{t_{\frac{1}{2}\ \text{desorption}}} \approx \frac{240}{6} = 40 \tag{4.4}$$

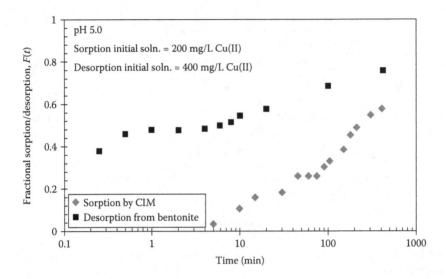

FIGURE 4.8 Comparison of fractional desorption rate of Cu(II) from the ion exchange sites of bentonite with the fractional sorption rate of dissolved Cu(II) from pure solution by the CIM.

Therefore, in any system where two solid phases coexist, the sorption of heavy metals from solution is most likely to be the rate-limiting step. Also, any practical application would involve cyclic use of the CIM. Since the heavy metal is removed from the CIM into an acid solution with each cycle, the CIM is totally free of heavy metals at the beginning of each cycle. This creates a maximum possible concentration gradient between the aqueous solution phase and the CIM phase, thus facilitating faster kinetics. Therefore, for the case of heavy-metal decontamination from ion exchange sites of clay, the kinetic limitations encountered would be due to the composition of the CIM and not from the desorption of heavy metals from the functional groups of the clay material.

4.2.3.4 Ion Selectivity of Bentonite

Figure 4.9 shows the results of Cu/Ca selectivity of bentonite at a pH of 4.5, and Table 4.2 details the coefficient and separation factor values for different compositions of the aqueous and CIM phases. In Figure 4.9, the aqueous-phase total cation concentration, $\{Ca^{2+}\} + \{Cu^{2+}\}$, remained almost the same. This proves that cation exchange was the major physicochemical process occurring.

For the Cu^{2+}/Ca^{2+} system, the separation factor of copper over calcium is given by

$$\alpha_{Cu/Ca} = \frac{y_{Cu}x_{Ca}}{x_{Cu}y_{Ca}} \tag{4.5}$$

where y represents the fraction of the element in the solid phase and x represents the fraction in the aqueous phase.

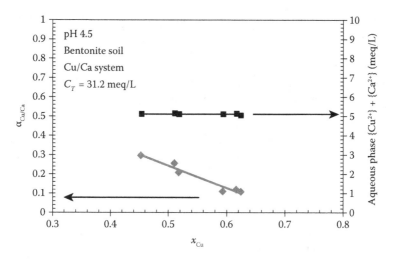

FIGURE 4.9   Copper–calcium selectivity of bentonite at pH 4.5.

TABLE 4.2
Cu/Na Binary System

y_{Cu} (dimensionless)	x_{Cu} (dimensionless)	$K_{Cu/Na}$ (g/L)	$\alpha_{Cu/Na}$ (dimensionless)
0.550	0.028	6389	42.43
0.600	0.031	7309	46.89
0.647	0.039	7283	45.16
0.739	0.044	11,722	61.52
0.780	0.064	10,009	51.85
0.794	0.087	7812	40.45
0.864	0.123	10,532	45.30

Average $K_{Cu/Na}$ = 8722 (g/L)
Average $\alpha_{Cu/Na}$ = 47.66 (dimensionless)

From Figure 4.9 it may be noted that:

1. $\alpha_{Cu/Ca}$ < 1.0, meaning thereby that the bentonite has more affinity for Ca^{2+}. This matches the observation reported in the literature[16,19,20] for strong-acid cation exchange resins. The primary reason attributed for calcium selectivity over copper is that Ca^{2+} ions cause less swelling of the resin than Cu^{2+}, or water uptake by the resin/clay is lower for Ca^{2+} form than that for Cu^{2+} form. This value of separation factor is in sharp contrast with $\alpha_{Cu/Ca} \approx 1000$ for IDA functionality chelating resin.[21]
2. $\alpha_{Cu/Ca}$ decreases with increasing x_{Cu}. This phenomenon has been observed in almost all strong-acid and strong-base ion exchangers and is attributed to the nonuniformity of ion exchange sites.[22]

3. The aqueous-phase total cation concentration, $\{Ca^{2+}\} + \{Cu^{2+}\}$, remained almost the same in each case, proving that ion exchange between Cu^{2+} and Ca^{2+} is the major physicochemical process occurring. A similar procedure was carried out to determine the selectivity of bentonite for the Cu/Na system. Since this is a heterovalent ion exchange system, the "selectivity coefficient" is thermodynamically a more rigorous parameter, and this value is reported in Table 4.2. The selectivity coefficient $K_{Cu/Na}$ can be expressed mathematically as

$$K_{Cu/Na} = \frac{y_{Cu} \cdot x_{Na}^2 \cdot C_T}{y_{Na}^2 \cdot x_{Cu} \cdot Q} \tag{4.6}$$

where C_T is the total aqueous-phase cation concentration, $\{Cu^{2+}\} + \{Na^+\}$, which ranged from 19 to 36 meq/L, and Q is the bentonite cation exchange capacity, which is equal to 0.5 meq/g.

The average separation factor values calculated for Cu/Ca and Cu/Na systems are $\alpha_{Cu/Ca} = 0.325$ and $\alpha_{Cu/Na} = 47.66$.

From the above observations, it can be concluded that the cation exchange behavior of smectite clay is similar to that of a strong-acid cation exchanger. Thus, any system containing bentonite loaded with heavy metals in its ion exchange sites and CIM can be mechanistically represented by the interaction of two cation exchangers, one with a selectivity preference that is the same as that of a strong-acid cation exchanger ($\alpha_{Cu/Na} \approx 50$) and the other with chelating functionality ($\alpha_{Cu/Na} \approx 10,000$).[9,21] For the case of Cu(II) desorption from ion exchange sites of bentonite by the addition of Na^+ in the aqueous phase, the exchange reactions involved can be summarized as follows:

$$\overline{(Z^-)_2 Cu^{2+}} + 2Na^+ \leftrightarrow \overline{2Z^- Na^+} + Cu^{2+}(aq) \tag{4.7}$$

$$\overline{R-N-(CH_2COO^- Na^+)_2} + Cu^{2+}(aq) \leftrightarrow \overline{R-N-(CH_2COO^-)_2 Cu^{2+}} + 2Na^+(aq) \tag{4.8}$$

Overall,

$$\overline{(Z^-)_2 Cu^{2+}} + \overline{R-N-(CH_2COO^- Na^+)_2} \leftrightarrow \overline{R-N-(CH_2COO^-)_2 Cu^{2+}} + \overline{2Z^- Na^+} \tag{4.9}$$

where Z and R represent the bentonite clay and the composite membrane, respectively. In order for the proposed process to succeed, the overall reaction (Equation 4.9) has to be thermodynamically favorable. Considering ideality, the equilibrium constant for such a reaction in terms of soil-phase and membrane-phase considerations is given as

$$K_{overall} = \frac{q_{Na}^Z \cdot q_{Cu}^R}{q_{Cu}^Z \cdot q_{Na}^R} \tag{4.10}$$

where the superscripts Z and R denote the soil phase and the membrane phase, respectively, while q_{Na} and q_{Ca} represent the sodium and copper concentrations in the corresponding solid phase. Multiplying both the numerator and denominator of Equation 4.10 by C_{Na}/C_{Cu} (C_i denotes the aqueous-phase concentrations of species i), we obtain

$$K_{overall} = \left[\frac{q_{Cu}^R / C_{Cu}}{q_{Na}^R / C_{Na}} \right]\left[\frac{q_{Na}^Z / C_{Na}}{q_{Cu}^Z / C_{Cu}} \right] = \frac{\alpha_{Cu/Na}^R}{\alpha_{Cu/Na}^Z} \tag{4.11}$$

Thus, $K_{overall}$ is the ratio of the Cu/Na separation factor between the composite membrane and the bentonite clay. As discussed earlier, due to the presence of the chelating functional group (IDA moiety), the dimensionless Cu/Na separation for the composite membrane is about two to three orders of magnitude greater compared to bentonite. As a result, the overall process is quite selective for decontamination of clays with high IEC. Thus, advantage is being taken of the fact that the bentonite clay ion exchanging functionality has much less selectivity for Cu(II) over Na (by introducing high concentration of Na in the aqueous phase), whereas the chelating membrane has very high affinity for Cu(II) over Na (due to which only Cu(II) is taken up by the membrane and concentrated in the regenerant solution).

4.2.3.5 Cyclic Process to Buffered Sludge at Alkaline pH with Aqueous-Phase Ligand Experimental Procedure

The solid phase in this case was prepared by mixing $CaCO_3$(s) and CuO in the ratio 10:1. The mass of the two compounds was such that $CaCO_3$(s) and $Cu(OH)_2$ were always the governing solid phases in the sludge. The solution volume was 200 ml. Six such solutions were prepared and a ligand (citrate or oxalate at concentrations of 0.01, 0.05, or 0.1 M) was added as sodium salt in each of them. Sorption/desorption steps were carried out for 10 cycles, with each cycle comprising a 4 h exhaustion period followed by a 1 h regeneration period. The choice of the aqueous-phase ligand is critical, and the effect of the ligand is discussed later. The two ligands (citrate and oxalate) were chosen because:

i. They form fairly strong complexes with most of the heavy metals.
ii. They are innocuous to the environment and also biodegradable and thus can be efficiently removed from the reactor after the last cycle.
iii. Most organic matter contains carboxylate functional groups; thus, citrate and oxalate can be used as surrogate representatives of complexing organic matter in the contaminated water.

Samples were taken from the sludge and the regenerant solution after the end of each cycle for Cu(II) and Ca(II) analysis.

Another experiment was conducted in a higher suspended solids content reactor. The solid phase in this experiment was prepared by mixing 5 g of $CaCO_3$(s), 45 g of fine sand (–200 mesh), 13.4g of $Na_2C_2O_4$, and 0.38g of CuO(s). One liter of DI water was added to this mixture, and the pH was adjusted to 9.0. The aqueous-phase oxalate concentration was 4000 mg/L, and <1% CuO(s) was present in the solid phase of the sludge. Sorption/desorption steps were subsequently carried out for 10 cycles, with Cu(II), Ca(II), and oxalate being analyzed after each cycle. Oxalate was analyzed using an ion chromatography with conductivity detector and standard carbonate/bicarbonate eluent.

4.2.3.6   Results

Figure 4.10 shows the Cu(II) uptake isotherms at two different oxalate concentrations. In both cases, there was no Cu(II) in the solid phase. To maintain this condition, the following procedure was adopted.

The desired oxalate solution (0.005 and 0.1 M) was prepared by adding a required mass of $Na_2C_2O_4$ to DI water. The pH was adjusted to 9.0 and the maximum $\{Cu_t\}$ that can be supported for each oxalate solution was experimentally determined. This was done by adding CuO(s) in mass much higher than that obtained by theoretical calculations based on Cu complexes with OH^- and $C_2O_4^{2-}$. The system was allowed to attain equilibrium, after which a sample was taken from the reactor, filtered, and analyzed for Cu(II). After determining the maximum $\{Cu_t\}$ for each case (8 mg/L for 0.005 M and 100 mg/L for 0.1 M), different amounts of CuO were added for each run after making sure that all the Cu added would remain in the aqueous phase.

It can be observed that the uptake of Cu(II) by the membrane depends only on the total aqueous-phase Cu(II) concentration; for $\{Cu_t\}$ <8 mg/L, the concentration of Na^+ or $C_2O_4^{2-}$ in the system makes practically no difference in Cu uptake even though they are 20 times higher in one case than the other.

Figures 4.11 through 4.13 compare the performance of the system with three ligands at the same concentration (0.05 M), that is, acetate, citrate, and oxalate.

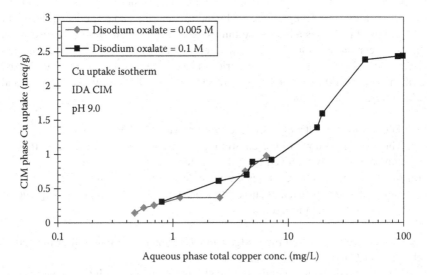

FIGURE 4.10   Copper (II) uptake isotherms at two oxalate concentrations.

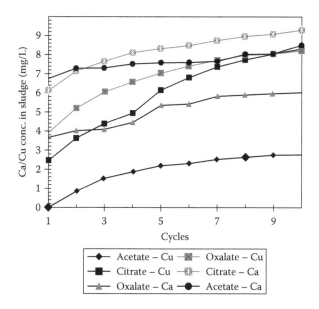

FIGURE 4.11   Aqueous-phase metals' concentration over number of cycles with different ligands.

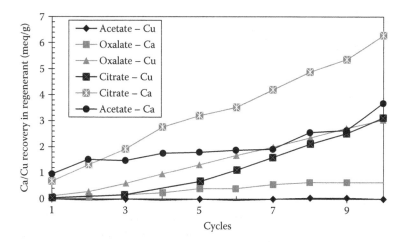

FIGURE 4.12   Metal uptake by IDA–CIM over number of cycles with different ligands.

It can be seen that citrate provided the highest sludge-phase concentrations of calcium while the acetate-added system showed the lowest reactor concentration of calcium. Because of this, the uptake of calcium by the CIM in descending order is citrate > oxalate > acetate. This is understandable from Table 4.3, which provides the stability constants of the ligands used in this study. Note that *N*-benzyl iminodiacetate (N-B-IDA) is the closest monomeric analog to the polystyrene–divinylbenzene IDA

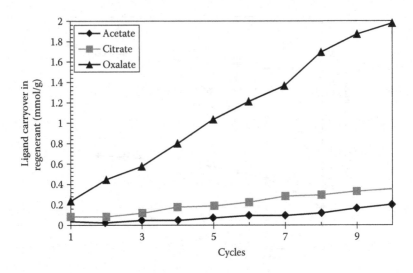

FIGURE 4.13   Ligand carryover in regenerant.

TABLE 4.3
Stability Constants for Cu^{2+}–Ligand and Ca^{2+}–Ligand Complexes^a

Ligand	Metal/Ligand	Log K Cu–Ligand	Log K Ca–Ligand
Oxalate	1:1	6.23	1.66
Oxalate	1:2	10.27	2.69
Citrate	1:1	5.90	3.50
Acetate	1:1	2.16	0.6
N-B-IDA	1:1	10.62	3.17
N-B-IDA	1:2	15.64	—

^{a} All values taken from Ref. 23.

functionality of the CIM. However, the case of copper uptake does not strictly follow the hierarchy as may be expected from the stability constants in Table 4.3. Acetate provided the lowest recovery of copper from the solid phase, but for the case of oxalate and citrate, an anomalous situation exists. Although aqueous-phase copper concentration is lower for oxalate systems as compared to citrate, uptake by CIM is comparable, especially in the latter cycles. Moreover, it may be observed that oxalate tended to be consistently taken up by the CIM, a phenomenon not observed for citrate or acetate. It may be pertinent to describe the experimental framework for oxalate system in more detail. In this case, to a 5% (w/v) sludge, fine sand (70%), Na$_2$C$_2$O$_4$ (21%), calcite (7.8%), and CuO (0.6%) were mixed. The sludge pH was maintained at 9.0, and CaC$_2$O$_4$ and Cu(OH)$_2$ were the controlling solid phases under these conditions. Thus, the amount of oxalate added was much more than 0.05 M. The oxalate added reacted with calcium to form calcium oxalate, a compound with low solubility. Precipitation of calcium oxalate reduced the aqueous-phase oxalate concentration to 0.05 M. After the formation of $CaC_2O_4(s)$, the cyclic process was started.

Figure 4.14 shows the recovery of Cu, Ca, and C_2O_4 in the regenerant solution for the oxalate system. Although Cu is present primarily as a $Cu–C_2O_4$ complex in the sludge phase, Cu recovery was significant and increased steadily with every cycle. Ca recovery was much lower and tended to approach an asymptotic concentration in the regenerant with an increase in the number of cycles. Dissolved sludge-phase concentrations of Cu and Ca remained fairly constant with the number of cycles (Figure 4.15), suggesting that they are controlled by the solubility products of the solid phases. Heavy-metal uptake by the CIM is substantial even when a free Cu ion, Cu^{2+}, is practically absent and most of the dissolved Cu exists primarily as anionic or neutral $Cu–C_2O_4$ complexes. Table 4.4 presents data about the computed percentage distribution of important Cu(II) species at three different oxalate concentrations based on stability constants data in the open literature.[23] Significant Cu uptake by the CIM under such conditions seems counterintuitive. Also, Figure 4.13 shows a consistently increasing oxalate recovery in the regenerant. Oxalate exists primarily as the divalent anion $C_2O_4^{2-}$ at a pH of 9.0 and should be rejected by the chelating cation exchanger according to the Donnan exclusion principle.

One possible explanation for this counterintuitive behavior could be that the suspended particles of the insoluble calcium oxalate and copper oxide were probably trapped in the large pores of the CIM and subsequently carried over to the regenerant phase, thus exhibiting significant copper and oxalate recovery. To eliminate such a possibility, another sorption isotherm was carried out in the absence of suspended solids at a pH of 9.0 and total aqueous-phase oxalate concentration of 4000 mg/L but at varying dissolved copper concentrations. Figure 4.16 shows Cu and C_2O_4 uptakes under these experimental conditions free of suspended solids. Note that the C_2O_4

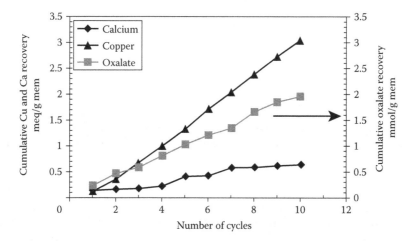

FIGURE 4.14   Cumulative copper, calcium, and oxalate recovery with number of cycles at pH 9.0.

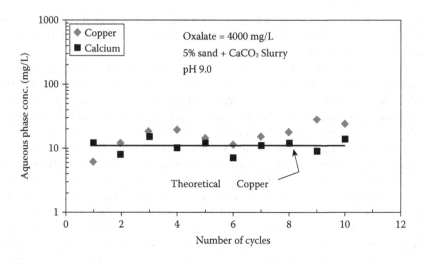

FIGURE 4.15   Dissolved copper and calcium concentrations during the recovery process at pH 9.0 for oxalate concentration of 4000 mg/L.

TABLE 4.4
Percentage Distribution of Various
Copper(II) Species

	Oxalate		
Species	0.005 M	0.045 M	0.1 M
Cu^{2+}	2.25E – 04	4.37E – 06	1.26E – 06
$Cu(OH)^+$	2.25E – 02	4.37E – 04	1.26E – 04
$Cu(OH)_2^0$	1.08E + 00	2.00E – 02	6.04E – 03
$Cu(OH)_3^-$	2.25E – 02	4.37E – 04	1.26E – 04
$Cu(C_2O_4)^0$	1.84E + 00	2.60E – 01	1.40E – 01
$Cu(C_2O_4)_2^{2-}$	9.70E + 01	9.97E + 01	9.99E + 01

uptake by the CIM is significant, and the Cu uptake increases with an increasing Cu concentration, a pattern very similar to those obtained in Figures 4.11 through 4.13. These observations strongly suggest that Cu and C_2O_4 are removed from the solution (or sludge) phase primarily through sorption processes. The following are identified as plausible binding mechanisms for C_2O_4 and Cu(II) onto the composite IDA membrane.

As already indicated, the neutral copper–oxalate complex $[Cu(C_2O_4)]^0$ was significantly present in the aqueous phase under experimental conditions, and only two of the four primary co-ordination numbers of Cu(II) are satisfied in this complex. Since these complexes are electrically neutral, they can permeate readily to the sorption sites containing nitrogen donor atoms, which can favorably satisfy the remaining coordination requirements of Cu(II). Figure 4.17 shows how electrically neutral 1:1

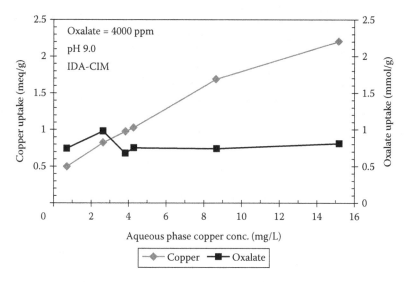

FIGURE 4.16   Equilibrium uptake of copper (II) and oxalate by the CIM at pH 9.0.

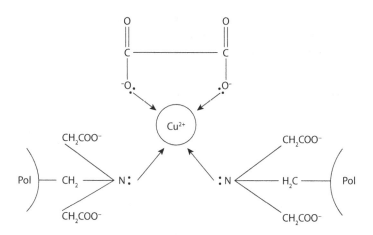

FIGURE 4.17   Sorption of neutral copper–oxalate complex onto chelating microbeads through formation of ternary complex.

copper–oxalate complexes can be bound to the neighboring nitrogen donor atoms of the IDA moieties through the formation of ternary complexes. This mode of sorption (i.e., formation of a ternary complex through a metal ligand interaction) is believed to be the primary pathway for the sorption of oxalate onto the CIM and subsequent carryover into the regenerant as exhibited in Figure 4.14.

Although free Cu ions (Cu^{2+}) were practically absent under the experimental conditions present in Figures 4.11, 4.12, 4.14, and 4.17, the Cu uptake was quite significant.

It is very likely that ligand substitution was a major mechanism by which Cu(II) was sorbed onto the chelating microbeads of the CIM for oxalate as aqueous-phase ligands. Table 4.3 shows 1:1 and 1:2 metal–ligand stability constants of Cu(II) with oxalate, citrate, acetate, and N-B-IDA, and the much higher ligand strength of IDA can be easily noted. Also, Cu(II) complexes are known to be labile. Therefore, the ligand substitution reaction, where oxalate in the aqueous phase is replaced by IDA in the exchanger phase, is favorable and results in an increased Cu(II) uptake as shown by the following reaction:

$$\left[Cu(C_2O_4)_2 \right]^{2-} + \overline{R-N-\left(CH_2COO-Na^+ \right)_2}$$

$$\leftrightarrow \overline{R-N-\left(CH_2COO^- \right)_2 \ Cu^{2+}} + 2Na^+ + 2C_2O_4^{2-}$$

(4.12)

Thus, ligand substitution and ternary complex formation are the major mechanisms for heavy-metal uptake in the case of oxalate as the aqueous-phase ligand. However, in the case of citrate or acetate, ligand substitution is the only mechanism for the transfer of copper from the sludge to the regenerant. This is verified by the almost total absence of citrate or acetate from the regenerant. The trace amounts of acetate and citrate noticed in the regenerant are attributed to their entrapment in the pores of the CIM. Therefore, Cu(II) uptake from an environment where citrate is the aqueous-phase ligand can be represented solely by the following equation:

$$\left[Cu(C_6O_7H_5) \right]^- + \overline{R-N-\left(CH_2COO-Na^+ \right)_2}$$

$$\leftrightarrow \overline{R-N-\left(CH_2COO^- \right)_2 \ Cu^{2+}} + 2Na^+ + 2C_6O_7H_5^{3-}$$

(4.13)

A comparative study of the three ligands reveals that:

1. Acetate is not an effective ligand for selective heavy-metal removal in a situation like the one simulated because it is too weak a ligand to impact aqueous-phase heavy-metal concentration, and thus is ineffective for subsequent CIM uptake.
2. The heavy-metal uptake profile is similar in the cases of citrate and oxalate, even though the aqueous-phase heavy-metal concentration of the heavy metal is higher for the citrate system. This is because the heavy metal is taken up by the CIM functionality by two parallel mechanisms (ligand substitution and ternary complex formation) in the case of oxalate addition, whereas ligand substitution is the only mechanism for heavy-metal uptake for the case of citrate addition.
3. Calcium uptake by the CIM is much higher in the case of citrate, as the aqueous-phase ligand, when compared with oxalate. This is because (i) the ligand citrate has a much higher affinity for calcium than oxalate and

(ii) calcium oxalate is sparingly soluble and thus most of the calcium remains in the solid phase. Since the free calcium concentration, $\{Ca^{2+}\}$, is lower because of CaC_2O_4 (s), the total calcium aqueous-phase concentration $\{Ca_t\}$ is also lower for oxalate relative to citrate, as can be noted from Figure 4.13. The major repercussion of the choice of ligand is in the suitability of the decontamination process. If one of the goals of solid-phase decontamination is recycle/reuse of the heavy metal, purity of the heavy metal in the acid regenerant stream would be important, and in such cases, oxalate would be a better candidate because of the lower calcium carryover. However, if the removal of heavy metal from the solid phase is the only goal, both oxalate and citrate would be suitable.

4. Oxalate sorption by the CIM is much higher than citrate for the reasons discussed earlier. Thus, if the ligand is undesirable in the regenerant, citrate would be a better choice than oxalate. Also, since citrate remains in the aqueous phase, stoichiometric amounts of it are needed in the reactor. In contrast, oxalate is removed from the liquid phase as a precipitate and thus its amount needed is much higher than stoichiometric.

From an application viewpoint, the concept is very important because it proves scientifically that heavy-metal-laden sludges with high buffer capacity can be decontaminated even at alkaline pH in the presence of organic ligands without the addition of acid to reduce pH; in fact, acid addition will not lower the pH because of high buffer capacity. The main concept employed here is to increase the aqueous-phase concentration of the heavy metal with minimal increase in the concentration of the competing cation in the solid phase. The stability constant values of the two ligands (aqueous ligand added to the solid-phase ligand of the CIM functionality) with the metals of interest play an important role in the suitability of the process.

Another methodology that can be employed is the use of an ion exchanger, which has very low affinity for Ca(II) as compared to that for heavy metal, say Cu(II). If one can identify a chelating functionality with affinity for heavy metals orders of magnitude higher than that for calcium, the need to add an aqueous-phase ligand would be obviated. In this regard, the work done with chelating polymers containing multiple nitrogen donor atoms[24–28] is significant. The authors have reported the presence of two chelating ion exchangers with pyridine base functionality with very high affinity for strong Lewis acids and very low affinity for "hard" acids such as Ca(II). If such functionality can be impregnated in the CIM, the process of cyclic extraction would become efficient.

4.2.4 CIM KINETICS

4.2.4.1 Experimental Protocol

Sorption kinetics of Cu^2, Pb^{2+}, and Ni^{2+} onto the CIM were studied at a pH of 3.0 or 5.0. Each experiment was performed in a 1.0 L baffled and cylindrical Plexiglas vessel and a computer-aided titrimeter to control the pH. The initial concentration of the heavy metal in the reactor was 200–230 mg/L added as its nitrate salt. About 0.5 g of previously conditioned CIM in basic form was added at the start following which approximately 2-mL volumes of samples were collected from the reactor at predetermined time intervals.

4.2.4.2   Results

Figure 4.18 shows almost identical uptake rates for Cu(II), Ni(II), and Pb(II) by the CIM, although they are not equally labile from a chemical reaction viewpoint. It is therefore likely that chemical reaction kinetics would not be the rate-limiting step. Previous investigations in this regard with spherical chelating exchangers have demonstrated intraparticle diffusion to be the most probable rate-controlling step.[29,30] Also, since the PTFE fibers do not sorb any metal ions, solute transport by surface diffusion is practically absent. However, a significantly different physical configuration of the CIM may introduce additional diffusional resistance within the CIM. As may be observed from the schematic diagram in Figure 4.19, a fairly stagnant pore liquid is present in the channels of the CIM between individual microbeads, and the solutes need to be transported through this pore liquid for sorption. This additional resistance is likely to retard the kinetic rate.

Figure 4.20 shows the results of a batch kinetic study (fractional metal uptake vs. time) comparing the parent chelating exchanger with the CIM under otherwise identical conditions. Fractional metal uptake, $F(t)$, is dimensionless and is defined as the ratio of the metal uptake $q(t)$ after time t and the metal uptake at equilibrium, q^0, that is, $F(t) = q(t)/q^0$. Although the parent chelating microbeads were bigger in average diameter (50–100 mesh) than the microbeads within the CIM, the copper uptake rate, as speculated, was slower with the CIM. In order to overcome the complexity arising due to the heterogeneity of the CIM (chelating microbeads randomly distributed in non-adsorbing Teflon fibers), a model was proposed[8–14] in which the thin-sheet–like CIM may be viewed as a flat plate containing a pseudo-homogeneous sorbent phase as shown in Figure 4.21.

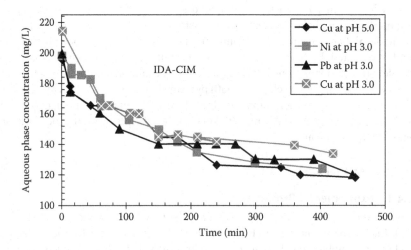

FIGURE 4.18   Plot of aqueous-phase metal concentration versus time during batch kinetic study of IDA–CIM.

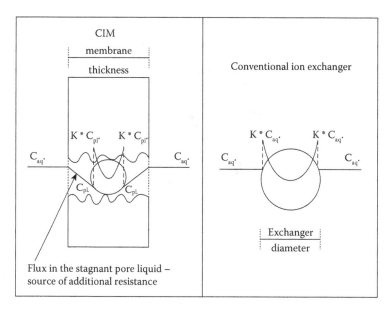

FIGURE 4.19   Schematic diagram showing the difference in the flux curves for a conventional (bead-shaped) ion exchanger and CIM.

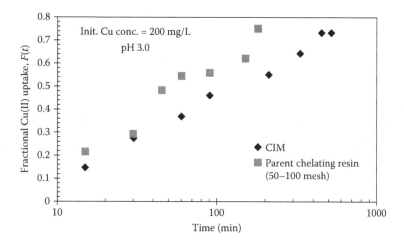

FIGURE 4.20   Copper uptake rates for the composite membrane versus the parent chelating exchanger under identical conditions.

Under the experimental conditions, it may be assumed that

1. The surface of the CIM is in equilibrium with the bulk of the liquid phase.
2. The total amount of solute in the solution and in the CIM sheet remains constant as the sorption process is carried out.
3. The solute (heavy metal) has high affinity toward the thin-sheet sorbent material.

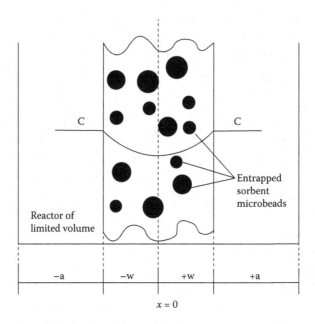

FIGURE 4.21 Schematic model showing sorption through an assumed flat plate with constant diffusivity from a reactor of limited volume.

This is a case of diffusion from a stirred solution with limited volume. The CIM is considered as a sheet of uniform material of thickness $2w$ placed in the solution containing the solute, which is allowed to diffuse into the sheet. The sheet occupies the space $-w \leq x \leq +w$, while the solution is of limited extent and occupies the space $-w - a \leq x \leq -w,\; w \leq x \leq +w + a$.

The concentration of the solute in the solution is always uniform and is initially C_0. The sheet is initially free from solute. Considering apparent metal–ion diffusivity within the CIM phase, the Partial Differential Equation (PDE) that must be solved in this case of metal uptake through a plane sheet (the thickness of the CIM) from a solution of limited volume is given as

$$\frac{\partial C}{\partial t} = \bar{D}\frac{\partial^2 C}{\partial x^2} \tag{4.14}$$

where x is the axial space coordinate in the direction of CIM thickness, and C is the concentration of the solute in the solution.

The Initial Condition (IC) of the above PDE is

$$C = 0, \quad -w < x < +W, \quad t = 0 \tag{4.15}$$

and the BC expresses the fact that the rate at which the solute leaves the solution is always equal to that at which it enters the sheet over the surface, $x = \pm w$. This condition is mathematically expressed as

$$\left(\frac{a}{K}\right)\frac{\partial C}{\partial t} = \pm\bar{D}\frac{\partial C}{\partial x}, \quad x = \pm w,\, t > 0 \tag{4.16}$$

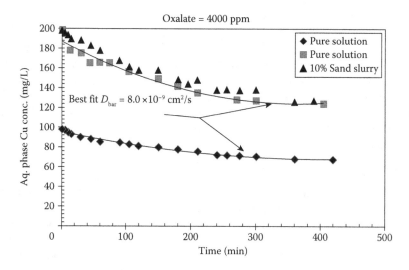

FIGURE 4.22 Comparison of kinetic model predictions (solid lines) with three independent experimental data sets.

where K is the partition factor between the CIM and the solution, that is, the concentration just within the sheet is K times that in the solution. An analytical solution of this problem is given as[31]

$$F(t) = \frac{q_t}{q^0} = 1 - \sum_{n=1}^{\infty} \frac{2\alpha(1+\alpha)}{1+\alpha+\alpha^2 q_n^2} \exp\left\{-\frac{\bar{D}q_n^2 t}{w^2}\right\} \qquad (4.17)$$

Figure 4.22 shows the results of three independent kinetic studies in which aqueous-phase copper concentrations are plotted against time; the two sets had different initial Cu concentrations (200 mg/L vs 100 mg/L) while the third one was a 10% fine sand slurry. The solid lines in Figure 4.22 indicate model predictions for apparent copper ion diffusivity of 8.0E – 9 cm²/s. Note that although the exact morphology, that is, distribution of voids and particles within the PTFE sheet, is not known, the suggested pseudo-homogeneous plane sheet model showed good agreement for the three sets of independent kinetic data produced with different initial concentrations.

4.3 ION EXCHANGE FIBER

4.3.1 BACKGROUND

A number of fiber-based ion exchange materials have been proposed. Chelating fibers with varying substrates (e.g., polymer fiber or natural fiber) and functional groups (e.g., amino, thio, oxo, carboxyl) have been developed.[32] IXF can be viewed as slender strands consisting of a polymeric backbone with functional groups covalently attached to the fiber surface. Compared to ion exchange resins, IXFs have a number of distinct advantages including faster kinetics (due to reduced length of ion transport) and the ability to be regenerated with benign reagents due to the placement of functional groups on or close to the surface.[33,34] IXF can also be used in reactors with highly suspended solids (not possible with IX resins) or be woven into filtration material or fabric. Other advantages include the ability to be easily compressed or loosened in a packed bed according to the operational requirements, and flexibility in usage for the removal of soluble pollutants. IXFs also have the unique advantage over other ion exchangers due to their ability to be processed into a wide range of diverse forms.[35,36]

A number of examples exist of the development and application of IXF.[33,34,37–39] IXFs have been created from a number of different polymers and utilized in various applications such as ultrafiltration, water treatment, and superabsorbent fibers.[32,36,40–42] PAN is considered one of the most important fiber-forming polymers due to its properties such as high strength, abrasion resistance, desirable chemical resistance, thermal stability, low flammability, and low cost. It has been used extensively in the form of membranes and micron-sized fibers for applications such as ultrafiltration,[43–48] water treatment,[35,39,45] and super absorbent fibers.[42] Shunkevich et al.[49] and Zhang et al.[50] reported the use of micron-sized PAN fibers as the basis for developing IXF.

The minimum diameter of aforementioned IXF reported is around 10 μm, which limits some of the advantages listed earlier. Submicron IXF, in theory, could significantly enhance the specific surface area, and therefore the number of functional groups on the surface. Electrospinning, a simple and cost-effective process, is a mature platform technology; excellent publications discuss its theory and applications.[32,51–55] It has been employed to generate submicron polymeric IXF[36,48,56–59] in the 40–2000 nm range.

4.3.2 IXF SYNTHESIS AND CHARACTERIZATION

PAN/N,N'-dimethylformamide solution with a concentration of 10 wt.% was prepared by stirring at a temperature of 70°C until a homogeneous solution was obtained. The viscosity of this solution was measured to be around 2280 cP. PAN fibers were electrospun at a constant flow rate of 2.4 mL/h that was maintained using a syringe pump (Harvard Apparatus, Holliston, Massachusetts). A voltage of 20 kV was applied between the needle and the collector plate using a high-voltage power supply (Gamma High Voltage Research, Ormond Beach, Florida), and the tip to collector distance was maintained at 21.6 cm. The electrospinning process was carried out at 20 ± 2°C. Electrospun fibers were collected on an aluminum plate placed on an insulated platform. Fiber collection was random and in the form of a non-woven mat. Electrospinning of PAN has been studied in detail[43,60–64] for various applications including carbon fiber preparation and environmental engineering applications, and the parameters used for electrospinning in the present study were chosen based on a review of the existing literature.

The electrospun PAN was treated chemically with an alkali to obtain an ion exchange material by converting the nitrile groups present in the native PAN to carboxylic acid groups. The material was hydrolyzed based on the conditions reported by Zhang et al.[47] Briefly, about 1 g of PAN was added to a 250 mL solution of 2 N NaOH or 2 N KOH prepared in DI water at different time and temperature conditions under stirring. After treatment, the material was allowed to cool and then was washed extensively with DI water. The functionalized PAN was dried in an oven at 40°C and put in an appropriate solution for determining the IEC. The experiments were replicated identically at least three times for IEC determination.

Figure 4.23a depicts an SEM image of the electrospun PAN fibers. The corresponding diameter distribution is shown in Figure 4.23b. Prior to functionalization, the SEM image reveals uniform and non-beaded fiber morphology. The average fiber diameter is ca. 720 nm and the diameter distribution shows a small tail, which is characteristic of the size distribution for electrospun fibers.[65]

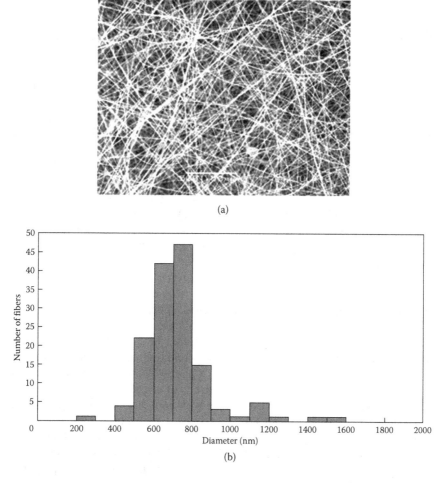

(a)

(b)

FIGURE 4.23 (a) SEM image of electrospun PAN fibers. (b) Diameter distribution of electrospun PAN fibers; average fiber diameter 720 nm.

TABLE 4.5

XPS Analysis of Electrospun PAN Fibers before (PAN) and after Hydrolysis (HPAN) with Sodium Hydroxide (Treated with 2 N NaOH at 70°C)

Atomic %	PAN	HPAN
Carbon	76.2	39.7
Oxygen	1.3	32.9
Nitrogen	22.6	0.7
Sodium		26.7
O/C ratio	0.017	0.829
N/C ratio	0.297	0.018

Table 4.5 shows the XPS data of atomic weight % of elements in electrospun PAN before and after hydrolysis with 2 N NaOH. The results depict an increase in the oxygen to carbon ratio from 0.017 to 0.829 and a decrease in the nitrogen to carbon ratio from 0.297 to 0.018. The nitrile group of PAN can be hydrolyzed to amide and carboxylic groups by reacting it with an alkali.[41,66] Litmanovich and Platé[67] hypothesized that, initially, hydrolysis results in ≈20% amidines situated in the nearest neighborhood of carboxylate groups, and after complete hydrolysis of the amidines, poly[(sodium acrylate)-*co*-acrylamide] is formed. However, we did not find any evidence of this reaction scheme. The almost complete absence of nitrogen in the hydrolyzed sample clearly indicates the absence of any amide. Figure 4.24 shows the XPS survey and elemental scan of alkali hydrolyzed electrospun PAN, where the binding energy values are reported for carbon 1s (285 eV), oxygen 1s (533 eV), nitrogen 1s (around 400 eV), and sodium 1s (1072 eV). It is clear that the nitrile bonds present in unfunctionalized PAN have been converted into carboxylic acid groups, with the hydrogen being replaced with sodium ions as the carboxylic functionality would remain mostly in the form of $-COO^-Na^+$.[43] It has been reported that PAN with a certain content of $-COOH$ swells easily when it is exposed to an aqueous medium.[68] The same phenomenon was observed after the hydrolysis reaction in the present experiments, further indicating the presence of carboxylic acid groups.

Figure 4.25 shows the FTIR scan for various electrospun PAN fibers before and after hydrolysis with NaOH and KOH. Here, the control refers to electrospun fibers treated with water under the same conditions to rule out any effect of water on the fibers at high temperature (70°C). The electrospun PAN fibers and the control PAN fibers show a characteristic peak around 2242 cm^{-1} that corresponds to the $-CN$ bond. The hydrolysis, with both NaOH and KOH, leads to the development of peaks around 1700 cm^{-1}, which is attributed to the C=O bond stretching for the carboxylic acid group and a broad peak around 3300 cm^{-1} that corresponds to the $-OH$ bond stretching present in the carboxylic acid.[43,46,69,70] It is evident from these results that the nitrile bonds present in PAN polymers are being hydrolyzed to carboxylic acid groups, which can be utilized for ion exchange applications.

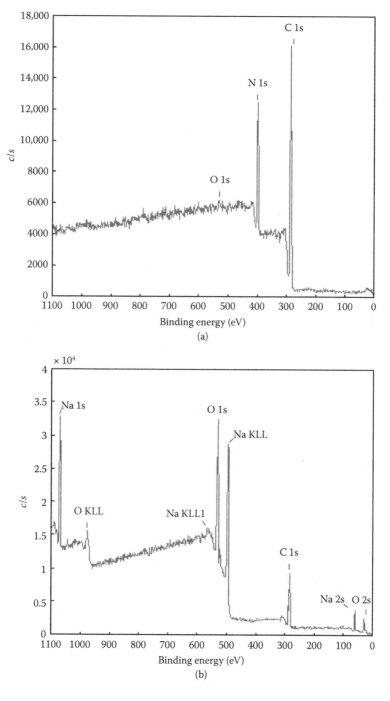

FIGURE 4.24 XPS analysis of electrospun PAN fibers: (a) before hydrolysis with sodium hydroxide and (b) after hydrolysis with sodium hydroxide.

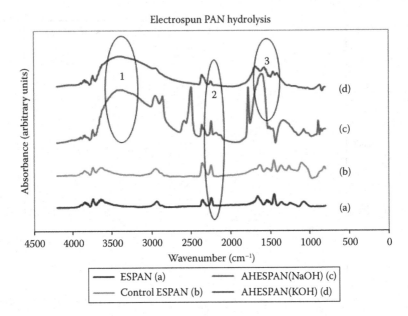

FIGURE 4.25 FTIR spectra for electrospun PAN fibers before and after hydrolysis with sodium hydroxide and potassium hydroxide. (1: –OH group, 2: –CN group, 3: –C=O group, ESPAN: electrospun PAN, AHESPAN: alkali-hydrolyzed electrospun PAN).

4.3.3 Optimization of Hydrolysis Conditions

Based on experiments done at different time and temperature conditions (data not shown), hydrolysis process conditions were optimized at 70°C and 30 min to obtain maximum IEC for electrospun fibers. Table 4.6 shows the IEC following hydrolysis with either NaOH or KOH for electrospun as well as micron-sized fibers. Control samples were prepared by treating electro-spun fibers with DI water (under the same reaction conditions utilized for functionalization by alkali hydrolysis) to determine if it resulted in the formation of hydrolyzed functional groups. Negligible IEC was found in the control samples, indicating that alkali treatment alone hydrolyzes nitrile groups of the PAN polymer and therefore is responsible for the formation of the carboxylate functional groups.

From Table 4.6, it is evident that treatment with KOH gives higher values as compared to treatment with NaOH. Zhang et al.[47] observed the difference in functionalization of PAN membranes due to the two different alkali species and attributed the difference (in the degree of hydrolysis) to the attack strength of the hydroxyl group. Young[71] posited this to the tendency to form ion pairs; that is, the tendency to form ion pairs is greatest in LiOH followed by NaOH, although experimental evidence was missing to support ion pair formation in KOH. In the present work, although the results are not significantly different, they are in agreement with the results reported by Zhang et al.[47]

TABLE 4.6

Effect of Different Alkali Species and Material Form of PAN Polymer on Ion Exchange Capacity

Experiment Set	Material	Reagent	Time (min)	Ion Exchange Capacity (meq/g dry membrane) (mean ± std dev)	Number of Repeats
Control	ES fibers	Water	30	0.03 ± 0.04	3
Hydrolyzed	ES fibers	2 N NaOH	30	1.96 ± 0.89	4
Hydrolyzed	ES fibers	2 N KOH	30	2.39 ± 0.87	4

Note: All experiments were conducted at 70°C.

4.3.4 HEAVY-METAL SELECTIVITY OF IXF

For the case of Cu^{2+}–Ca^{2+} exchange described in the earlier section, the ion exchange reaction may be written as

$$\overline{(R^+COO^-)_2Ca^{2+}} + Cu^{2+} \leftrightarrow \overline{(R^+COO^-)_2Cu^{2+}} + Ca^{2+} \qquad (4.18)$$

A modified form of the equilibrium relationship for this reaction is given by the separation factor, which is a dimensionless measure of relative selectivity between two competing ions and, in this case, equal to the ratio of distribution coefficient of Cu(II) concentration between the exchanger phase and the aqueous phase to that of calcium ion and is given as

$$\alpha_{Cu/Ca} = \frac{\left[\overline{(R^+COO^-)_2Cu^{2+}}\right]\left[Ca^{2+}\right]}{\left[\overline{(R^+COO^-)_2Ca^{2+}}\right]Cu^{2+}} \qquad (4.19)$$

For this reaction, the Coulombic and hydrophobic interactions play an insignificant role since both the ions are divalent. But the Lewis acid–base interaction plays a critical role since the binding constant for Cu with the closest ligand analog in the aqueous phase (acetate) is almost 10 times higher than that with Ca – log CaL binding constant = 1.2 while log CuL binding constant = 2.2, and $CuL_2 = 2.6$.[72] Figure 4.26 shows the profile of $\alpha_{Cu/Ca}$ for the hydrolyzed electrospun fiber. The results match data obtained by Roy[21] and SenGupta et al.[25] for carboxylate resin. Sengupta and SenGupta[73] have demonstrated a linear relationship between an experimentally determined metal/calcium separation factor for a commercial carboxylate resin (DP-1, no commercial endorsement implied) and aqueous-phase metal-acetate stability constant values. The data available in Figure 4.26 also agree with theoretical explanations of the reduction of the heavy-metal separation factor value with increase in the aqueous-phase heavy-metal concentration.[16,74,75] Thus, it can be stated that equilibrium properties of the carboxylate electrospun fiber are no different from the carboxylate ion exchange resin.

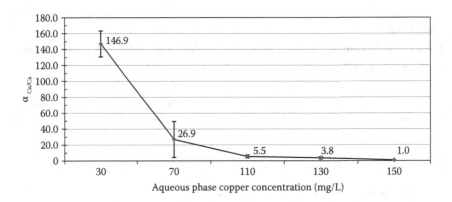

FIGURE 4.26 Plot of selectivity factor of copper over calcium for hydrolyzed electrospun PAN fibers.

4.3.5 IXF KINETICS

Figure 4.27 shows the profile of fractional uptake of Cu by IXF and commercially available resin beads of the same functional group (carboxylate) versus time of contact. Fractional uptake, $F(t) = M_t/M_\infty$, is the ratio of mass of Cu transferred into the exchanger from the aqueous phase at a particular time to the equilibrium mass uptake, which was experimentally confirmed to be achieved after 24 h of contact. A detailed discussion of a kinetic model of ion uptake is outside the central objective of this chapter. In IXF, particle diffusion is usually considered to be the rate-limiting step[16,76,77] (as opposed to liquid film diffusion or chemical reaction). For a two-dimensional analysis of the fiber, the governing equation to model the ion exchange kinetics can be written as[78]

$$\frac{\partial \bar{c}}{\partial t} = \bar{D} \frac{\partial^2 \bar{c}}{\partial r^2} \tag{4.20}$$

where $\bar{D}$ is the diffusivity of fiber, $\bar{c}$ the ion concentration in the fiber as a function of space and time, $R_0 - r_0$ the depth of penetration of counterion from the bulk solution, and L the length of the fiber.

The initial and boundary conditions for the above equation are

$$\text{At } t = 0, \quad \bar{c}(r, 0) = 0 \tag{4.21}$$

$$\text{At } r = r_0, \quad \frac{\partial \bar{c}(r_0, t)}{\partial r} = 0 \tag{4.22}$$

$$\text{At } r = R_0, \quad \bar{D} \frac{\partial \bar{c}(R_0, t)}{\partial r} = k_f \left(c(t) - \frac{\bar{c}(R_0, t)}{K} \right) \tag{4.23}$$

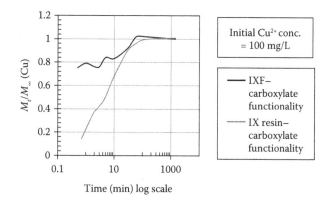

FIGURE 4.27 A comparison of kinetic uptake of copper for hydrolyzed IXF and ion exchange beads with carboxylate functionality. (M_t/M_∞ represents the uptake of copper at a specified time as compared with equilibrium copper uptake at 24 h).

where k_f is the mass transfer coefficient of the film at the interphase of the solid–liquid boundary, $c(t)$ the bulk ion concentration, $\bar{c}$ (R_0, t) the ion concentration at $r = R_0$ at the surface of the fiber, and K the partition coefficient.

From this model, it is clear that the rate of uptake is inversely proportional to the diameter of the fiber. It is easily noticeable from Figure 4.27 that at contact time <10 minutes—the range that is most likely to be the case for a real packed-bed operation—the IXFs show a much faster rate of uptake than ion exchange resin beads. This can be easily attributed to the difference in their size, as discussed earlier.

4.4 CONCLUSIONS

The presence of toxic and hazardous wastes with high suspended solids containing heavy metal(s) as the target contaminant is a widespread problem. In cases where the sludges contain heavy-metal precipitates along with other bulk nontoxic but chemically interacting solids, the selective removal of the target heavy metal(s) would be an ideal solution but also a challenging separation problem. Chelating ion exchangers can selectively remove dissolved heavy metals but their morphology (spherical beads of diameter <2 mm) does not permit usage in a sludge/slurry scenario. We have identified, developed, and studied extensively the suitability of using chelating ion exchangers but in two forms (composite membrane and IXF) such that they can be used in a reactor containing high concentrations of suspended solids. Moreover, by modifying the chemical constituents in the reactor—primarily by introducing innocuous, nontoxic species, one can increase the efficiency of the process causing selective removal of the heavy metal from the sludge to the composite membrane—all in a single reactor. The process is successful in concentrating the heavy metals in an acid solution, thereby providing an opportunity leading to heavy-metal recovery. Extensive laboratory experiments were carried out to investigate some key aspects of the proposed process. Primary conclusions of the study can be summarized as follows:

- The composite membrane was able to remove copper(II) from a synthesized sludge with a high buffer capacity by the introduction of an aqueous-phase ligand. Comparative studies showed that of the two ligands (citrate and oxalate) studied, citrate was more effective in heavy-metal removal but it also demonstrated more competing ion uptake in the regenerant solution, while oxalate had a tendency to be sorbed by the ion exchange membrane.
- The composite membrane was capable of removing Cu(II) from Cu(II)-loaded bentonite clay. Copper ions were selectively transferred from the ion exchange sites of bentonite clay to the ion exchange sites of the composite membrane.
- The physical composition of the composite membrane has a significant effect on the kinetics of the process. The membrane is composed of ion exchanger beads but stagnant pores inside the membrane present an extra diffusional resistance to the path of dissolved heavy-metal ions involved in the exchange reaction. This phenomenon is primarily responsible for our observation that the kinetics of sorption of heavy metal by the membrane is much slower than that of desorption (of the heavy metal) from bentonite clay; therefore, sorption by the composite membrane is the rate-limiting kinetic step. This observation suggests that the process efficiency can be improved by eliminating or reducing the intra-membrane stagnant pores in the composite membrane.
- The composite membrane was found to be compatible with the sludge medium and durable in the sludge/slurry reactor environment over a large number of cycles. Sorption properties of the composite membrane and its durability were unaffected by the concentration of the bulk suspended solids in the reactor.
- Submicron-sized PAN fibers were prepared by the simple technique of electrospinning and were subsequently functionalized via alkali hydrolyzation to yield weak acid IXF. Optimum hydrolysis conditions provided an IEC of 2.39 ± 0.87 meq/g. IEC was primarily attributed to conversion of nitrile groups to carboxylic acid groups on the PAN fibers; this was supported by XPS and FTIR spectra. The principles outlined herein can be utilized to generate submicron IXF with choice of diameter in the range of 200–900 nm.

APPENDIX 4.A

Let us take the generic case of a heavy-metal precipitate, say $MeCO_3(s)$. If it is the only solid phase, and $CO_2(g)$ is introduced to lower the pH and concomitantly increase the aqueous phase $\{Me^{2+}\}$, the relevant equations are

$$MeCO_3(s) \rightleftharpoons Me^{2+} + CO_3^{2-}; K_{sp} \qquad (4.24)$$

$$CO_3^{2-} + H^+ \rightleftharpoons HCO_3^-; \, 1/K_{a,2} \tag{4.25}$$

$$H_2CO_3^* \rightleftharpoons H^+ + HCO_3^-; \, K_{a,1} \tag{4.26}$$

$$CO_2(g) + H_2O \rightleftharpoons H_2CO_3^*; \, K_H \tag{4.27}$$

Adding Equations 4.24 through 4.27, we get

$$MeCO_3(s) + CO_2(g) + H_2O \rightleftharpoons Me^{2+} + 2HCO_3^- \, K_{eq,1} = \frac{K_{sp} \cdot K_{a,1} \cdot K_H}{K_{a,2}} = \frac{\left[Me^{2+}\right][HCO_3^-]^2}{pCO_2(g)} \tag{4.28}$$

The electroneutrality equation is

$$[H^+] + 2[Me^{2+}] = [OH^-] + [HCO_3^-] + 2[CO_3^{2-}] \tag{4.29}$$

For pH ϵ (4.3, 8.3), [H$^+$], [OH$^-$], and [CO$_3^{2-}$] can be ignored, and the simplified Equation 4.29 may be written as

$$2[Me^{2+}] = [HCO_3^-] \tag{4.30}$$

Equation 4.28 may be rewritten as

$$K_{eq,1} = \frac{K_{sp} \cdot K_{a,1} \cdot K_H}{K_{a,2}} = \frac{\left[Me^{2+}\right]\{2[Me^{2+}]\}^2}{pCO_2(g)}$$

$$[Me^{2+}]^3 = \frac{pCO_2 K_{sp} \cdot K_{a,1} \cdot K_H}{K_{a,2}} \tag{4.31}$$

$$[Me^{2+}] = \text{constant} * [pCO_2(g)]^{1/3}$$

Thus, a plot of p[Me^{2+}] versus log[pCO$_2$(g)] would have a slope of (1/3). Also, from Equation 4.26,

$$K_{a,1} = \frac{\left[Me^{2+}\right]\left[HCO_3^-\right]}{\left[H_2CO_3^*\right]} = \frac{\left[Me^{2+}\right]\left[HCO_3^-\right]}{K_H \cdot pCO_2(g)}$$

Thus,

$$[H^+] = \frac{pCO_{2(g)} \cdot K_{a,1} \cdot K_H}{\left[HCO_3^-\right]} \tag{4.32}$$

Substituting [HCO_3^-] from Equation 4.30 into Equation 4.32,

$$[H^+] = \frac{pCO_{2(g)} \cdot K_{a,1} \cdot K_H}{2 \left[Me^{2+} \right]}$$

(4.33)

Substituting [Me^{2+}] from Equation 4.31,

$$[H^+] = \frac{pCO_{2(g)} \cdot K_{a,1} \cdot K_H}{2 \left\{ \dfrac{pCO_2 K_{sp} \cdot K_{a,1} \cdot K_H}{K_{a,2}} \right\}^{\frac{1}{3}}}$$

$[H^+]$ = constant $*$ $pCO_2^{2/3}$.
Thus, a plot of $pCO_2(g)$ versus pH would have a slope of 2/3.
Let us substitute $MeCO_3$ with $PbCO_3$.
Figure 4.1a shows a profile of pPb^{2+} and pH as a function of partial pressure of $CO_2(g)$. It can be seen that [Pb^{2+}] can be influenced by $ppCO_2$.
Figure 4.1b shows that profile of pPb^{2+} and pH versus $pCO_2(g)$, and a slope of 1/3 for pPb^{2+} and 2/3 for pH is confirmed.

APPENDIX 4.B

Heterogeneous equilibria—More than one solid phase is present, say $PbCO_3(s)$ and $CaCO_3(s)$ are present together.
    Relevant equations:

$$PbCO_3(s) \leftrightharpoons Pb^{2+} + CO_3^{2-}; \; K_{sp1}$$

(4.34)

$$CaCO_3(s) \leftrightharpoons Ca^{2+} + CO_3^{2-}; \; K_{sp2}$$

(4.35)

$$2CO_3^{2-} + 2H^+ \leftrightharpoons 2HCO_3^-; (1/K_{a,2})^2$$

(4.36)

$$2H_2CO_3^* \leftrightharpoons 2H^+ + 2HCO_3^-; (K_{a,1})^2$$

(4.37)

$$2CO_2(g) + 2H_2O \leftrightharpoons 2H_2CO_3^* \; (K_H)^2$$

(4.38)

Adding Equations 4.34 through 4.38,

$$PbCO_3(s) + CaCO_3(s) + 2CO_2(g) \leftrightharpoons Pb^{2+} + Ca^{2+} + 4HCO_3^-$$

(4.39)

$$K_{eq,2} = \frac{K_{sp1}K_{sp2}(K_{a,1})^2(K_H)^2}{(K_{a,2})^2} = \{[Pb^{2+}][Ca^{2+}]\} * \frac{[HCO_3^-]^4}{[pCO_2(g)]^2}$$

The electroneutrality equation is

$$[H^+] + 2[Pb^{2+}] + 2[Ca^{2+}] = [OH^-] + 2[CO_3^-] + [HCO_3^-] \qquad (4.40)$$

If pH ϵ (4.3, 8.3), $[H^+]$, $[OH^-]$, and $[CO_3^{2-}]$ can be ignored, and Equation 4.40 may be rewritten as

$$2[Pb^{2+}] + 2[Ca^{2+}] = [HCO_3^-] \qquad (4.41)$$

Also,

$$K_{sp1} = [Pb^{2+}][CO_3^{2-}]$$

Thus,

$$[Pb^{2+}] = \frac{K_{sp1}}{[CO_3^{2-}]}$$

$$K_{sp2} = [Ca^{2+}][CO_3^{2-}]$$

$$[Ca^{2+}] = \frac{K_{sp2}}{[CO_3^{2-}]}$$

$$\frac{[Pb^{2+}]}{[Ca^{2+}]} = \frac{K_{sp1}}{K_{sp2}} \qquad (4.42)$$

$$[Ca^{2+}] = [Pb^{2+}] \frac{K_{sp2}}{K_{sp1}}$$

Substituting [Ca²⁺] from Equation 4.42 into Equation 4.41,

$$2[Pb^{2+}] + 2\left\{[Pb^{2+}]\frac{K_{sp2}}{K_{sp1}}\right\} = [HCO_3^-]$$

$$2[Pb^{2+}]\left\{1 + \frac{K_{sp2}}{K_{sp1}}\right\} = [HCO_3^-] \qquad (4.43)$$

Substituting $[HCO_3^-]$ from Equation 4.43 and $[Ca^{2+}]$ from Equation 4.42 into Equation 4.39:

$$K_{eq,2} \frac{[Pb^{2+}]}{[pCO_2(g)]^2}[Pb^{2+}]\frac{K_{sp2}}{K_{sp1}} \cdot \left\{ 2[Pb^{2+}]\left(1+\frac{K_{sp2}}{K_{sp1}}\right)\right\}^4$$

$$= 16\frac{[Pb^{2+}]^6}{[pCO_2(g)]^2}\frac{K_{sp2}}{K_{sp2}} \cdot \left\{\left(1+\frac{K_{sp2}}{K_{sp1}}\right)\right\}^4 \tag{4.44}$$

For $CaCO_3(s)$ and $PbCO_3(s)$, $K_{sp2} \gg K_{sp1}$

$$K_{eq2} = 16\frac{[Pb^{2+}]^6}{[pCO_2(g)]^2}\frac{K_{sp2}}{K_{sp1}} \cdot \left(\frac{K_{sp2}}{K_{sp1}}\right)^4 \tag{4.45}$$

Again,

$$K_{eq2} = K_{sp1}K_{sp2}\left(\frac{K_{a,1}K_H}{K_{a,2}}\right)^2$$

Thus,

$$[Pb^{2+}]^6 = \frac{[pCO_2(g)]^2}{16} \cdot \left(\frac{K_{sp2}}{K_{sp1}}\right)^5 \cdot K_{eq2}$$

$$= \frac{[pCO_2(g)]^2}{16}\left(\frac{K_{sp2}}{K_{sp1}}\right)^5 K_{sp1}K_{sp2}\left(\frac{K_{a,1}K_H}{K_{a,2}}\right)^2$$

$$= [pCO_2(g)]^2 \cdot \text{constant} \cdot \frac{K_{sp1}^6}{K_{sp2}^4} \tag{4.46}$$

For a constant $pCO_2(g)$,

$$[Pb^{2+}] \propto K_{sp1}$$

$$\propto K_{sp2}^{-2/3}$$

Thus, we see from Appendix 4.A,

$$[Pb^{2+}] \propto (K_{sp1})^{1/3}$$

whereas in this appendix,

$$[Pb^{2+}] \propto K_{sp1}$$

Moreover, from this appendix we also note that

$$[Pb^{2+}] \, \alpha \, K_{sp2}^{-2/3}$$

that is, the solubility of $PbCO_3$ is also dependent on the solubility constant of $CaCO_3$. Furthermore, from Equation 4.41,

$$2[Pb^{2+}] + 2[Ca^{2+}] = [HCO_3^-]$$

and from Equation 4.42,

$$[Pb^{2+}] = [Ca^{2+}] \frac{K_{sp1}}{K_{sp2}}$$

Thus,

$$2\left\{ [Ca^{2+}] \frac{K_{sp1}}{K_{sp2}} \right\} + 2[Ca^{2+}] = [HCO_3^-] \tag{4.47}$$

since $K_{sp2} \gg K_{sp1}$.

The above equation can be simplified to

$$2[Ca^{2+}] = [HCO_3^-] \tag{4.48}$$

Substituting $[HCO_3^-]$ from Equation 4.48 into Equation 4.35,

$$K_{eq2} = K_{sp1}K_{sp2}\left(\frac{K_{a,1}K_H}{K_{a,2}} \right)^2 = \frac{Ca^{2+}}{[pCO_2(g)]^2} \frac{K_{sp1}}{K_{sp2}} * [Ca^{2+}] * \{2[Ca^{2+}]\}^4$$

$$= 16 \cdot \frac{[Ca^{2+}]^6}{[pCO_2(g)]^2} \frac{K_{sp1}}{K_{sp2}} = K_{sp1}K_{sp2}\left(\frac{K_{a,1}K_H}{K_{a,2}} \right)^2$$

Thus,

$$[Ca^{2+}]^6 = \frac{[pCO_2(g)]^2}{16}(K_{sp2})^2\left(\frac{K_{a,1}K_H}{K_{a,2}} \right)^2$$

$$[Ca^{2+}] \, \alpha \, [pCO_2(g)]^{1/3}$$

$$\alpha \, K_{sp2}^{1/3}$$

Thus, the solubility of Ca^{2+} is not dependent on the solubility product of $PbCO_3$, but the opposite does not hold.

Also, since

$$[Pb^{2+}] = [Ca^{2+}] \frac{K_{sp1}}{K_{sp2}}$$

$[Ca^{2+}]$ will be $\gg [Pb^{2+}]$ at all pCO_2. This is confirmed by Figure 4.2a, which plots pPb^{2+}, pCa^{2+}, and pH versus $ppCO_2$.

Furthermore,

$$\frac{\left[H^+\right]\left[HCO_3^-\right]}{\left[H_2CO_3^*\right]} = K_{a,1}$$

$$= \frac{\left[H^+\right]\left[HCO_3^-\right]}{K_H \cdot \left[pCO_2(g)\right]}$$

From Equation 4.48,

$$2[Ca^{2+}] = HCO_3^-$$

and

$$[Ca^{2+}] = [pCO_2]^{1/3} \cdot constant$$

$$HCO_3^- = [pCO_2]^{1/3} \cdot constant/2$$

$$\left[H^+\right] = \frac{K_{a,1} K_H pCO_{2(g)}}{\left[HCO_3^-\right]}$$

$$= \frac{constant \left[pCO_{2(g)}\right]}{constant \, [pCO_{2(g)}]^{1/3}}$$

$$= constant \, [pCO_2(g)]^{2/3}$$

Figure 4.2b plots pPb^{2+}, pCa^{2+}, and pH versus $pCO_2(g)$ and it may be noted that the slope of pPb^{2+} and Ca^{2+} is 1/3 while that of pH is 2/3.

APPENDIX 4.C

Consider the solid phases $PbCrO_4$, $PbCO_3$, and $CaCO_3$ present in an aqueous solution. For a known $ppCO_2(g)$, and therefore known $\left[H_2CO_3^*\right]$, the unknown species present in the system are

$$Pb^{2+}, Ca^{2+}, H^+, OH^-, HCO_3^-, CO_3^{2-}, HCrO_4^-, CrO_4^{2-}, \text{ and } Cr_2O_7^{2-}$$

Since the system is open to the atmosphere,

$$H_2O + CO_2(g) \rightleftharpoons H_2CO_3^*$$

$$[H_2CO_3^*] = K_H * ppCO_2(g) \tag{4.49}$$

$$PbCrO_4 \rightleftharpoons Pb^{2+} + CrO_4^{2-}$$

$$K_{sp,1} = [Pb^{2+}] * [CrO_4^{2-}] \tag{4.50}$$

$$PbCO_3 \rightleftharpoons Pb^{2+} + CO_3^{2-}$$

$$K_{sp,2} = [Pb^{2+}] * [CO_3^{2-}] \tag{4.51}$$

$$CaCO_3 \rightleftharpoons Ca^{2+} + CO_3^{2-}$$

$$K_{sp,3} = [Ca^{2+}] * [CO_3^{2-}] \tag{4.52}$$

$$H_2CO_3^* \rightleftharpoons H^+ + HCO_3^-$$

$$K_{a,1} = \frac{\left[H^+\right]\left[HCO_3^-\right]}{\left[H_2CO_3^*\right]} = \frac{\left[H^+\right]\left[HCO_3^-\right]}{K_H * ppCO_2(g)} \tag{4.53}$$

$$K_{a,1} * K_H = \frac{\left[H^+\right]\left[HCO_3^-\right]}{ppCO_2(g)} \tag{4.54}$$

$$HCO_3^- \rightleftharpoons H^+ + CO_3^{2-}$$

$$K_{a,2} = \frac{\left[H^+\right]\left[CO_3^{2-}\right]}{\left[HCO_3^-\right]} \tag{4.55}$$

$$HCrO_4^- \rightleftharpoons H^+ + CrO_4^{2-}$$

$$K_{a,3} = \frac{\left[H^+\right]\left[CrO_4^{2-}\right]}{\left[HCrO_4^-\right]} \tag{4.56}$$

$$H_2O \rightleftharpoons H^+ + OH^-$$

$$K_w = [H^+] * [OH^-] \tag{4.57}$$

$$2CrO_4^{2-} + 2H^+ \rightleftharpoons 2Cr_2O_7^{2-} + H_2O$$

$$K_{a,4} = \frac{\left[Cr_2O_7^{2-} \right]^2}{\left[H^+ \right]^2 \left[CrO_4^{2-} \right]^2} \tag{4.58}$$

The electroneutrality equation is given as

$$2[Pb^{2+}] + 2[Ca^{2+}] + [H^+] = [OH^-] + [HCO_3^-] + [HCrO_4^-] + 2[CrO_4^{2-}]$$

$$+ 2[Cr_2O_7^{2-}] + 2[CO_3^{2-}] \tag{4.59}$$

The above equations are solved in MINEQL+ 4.6, and the result is shown in Figure 4.3, where

$$[Cr_t] = [HCrO_4^-] + [CrO_4^{2-}] + 2[Cr_2O_7^{2-}]$$

REFERENCES

1. United States Environmental Protection Agency. *SW-846: Test Methods for Evaluating Solid Waste, Physical/Chemical Method.* http://www.epa.gov/wastes/hazard/test-methods/sw846/online/index.htm (accessed March 12, 2015).
2. McLaughlin, M. J., Zarcinas, B. A., Stevens, D. P., Cook, N. Soil testing for heavy metals. *Commun. Soil Sci. Plant Anal.* **2000**, *31*, 1661–1700.
3. McLaughlin, M. J., Hamon, R. E., McLaren, R. G., Speir, T. W., Rogers, S. L. Review: A bioavailability-based rationale for controlling metal and metalloid contamination of agricultural land in Australia and New Zealand. *Aust. J. Soil Res.* **2000**, *38*, 1037–1086.
4. Ling, W., Shen, Q., Gao, Y., Gu, X., Yang, Z. Use of bentonite to control the release of copper from contaminated soils. *Aust. J. Soil Res.* **2007**, *45*, 618–623.
5. Wuana, R. A., Okieimen, F. E. Heavy metals in contaminated soils: A review of sources, chemistry, risks and best available strategies for remediation. *ISRN Ecol.* **2011**, *2011*, 402647, 20 pp. doi: 10.5402/2011/402647.
6. U.S. EPA. Clean Water Act, sec. 503, vol. 58, no. 32. U.S. Environmental Protection Agency, Washington, DC, 1993.
7. *MINEQL+ 4.6.* Environmental Research Software. Hallowell, ME.
8. Sengupta, S., SenGupta, A. K. Characterizing a new class of sorptive/desorptive ion exchange membranes for decontamination of heavy-metal-laden sludges. *Environ. Sci. Technol.* **1997**, *27*, 2133–2140.
9. Sengupta, S. *A new separation and decontamination technique for heavy-metal-laden sludges using sorptive/desorptive ion-exchange membranes.* PhD Dissertation, Lehigh University, Easton, PA, 1994.
10. Sengupta, S., SenGupta, A. K. Solid phase heavy metal separation using composite ion-exchange membranes. *Hazard. Waste Hazard. Mater.* **1996**, *13*, 245–263.
11. Sengupta, S., SenGupta, A. K. Heavy-metal separation from sludge using chelating ion exchangers with nontraditional morphology. *React. Funct. Polym.* **1997**, *35*, 111–134.
12. Sengupta, S. Electro-partitioning with composite ion-exchange material: An innovative in-situ heavy metal decontamination process. *React. Funct. Polym.* **1999**, *40*, 263–273.

13. Sengupta, S., SenGupta, A. K. Chelating ion-exchangers embedded in PTFE for decontamination of heavy-metal-laden sludges and soils. *Colloids Surf. A.* **2001**, *191*, 79–95.
14. Jassal, M., Bhowmick, S., Sengupta, S., Patra, P. K., Walker, D. I. Hydrolyzed poly(acrylonitrile) electrospun ion-exchange fibers. *Environ. Eng. Sci.* **2014**, *31*, 288–299.
15. Errede, A. A., Stoesz, L. N., Sirvio, J. Swelling of particulate polymers enmeshed in poly(tetrafluoroethylene). *J. Appl. Polym. Sci.* **1986**, *31*, 2721–2737.
16. Helfferich, F. G. *Ion Exchange*. McGraw Hill, New York, 1962.
17. SenGupta, A. K., Shi, B. Selective alum recovery from clarifier sludge. *J. AWWA.* **1992**, *84*, 96–103.
18. Grim, R. E. *Clay Mineralogy*. McGraw Hill, New York, 1968.
19. Kunin, R. *Ion Exchange Resins*. Wiley, New York, 1958.
20. Bonner, O. D., Livingston, F. L. Cation exchange equilibria involving some divalent ions. *J. Phys. Chem.* **1956**, *60*, 530–532.
21. Roy, T. K. *Chelating polymers: Their properties and applications in relation to removal, recovery and separation of toxic metal cations*. M.S. Thesis, Lehigh University, Easton, PA, 1989.
22. Reichenberg, D., McCauley, D. J. Properties of ion-exchange resins in relation to their structure. Part VII: Cation-exchange equilibria on sulfonated polystyrene resins of varying degrees of crosslinking. *J. Chem. Soc. 1955, III, 2741–2749.*
23. Smith, R. M., Martell, A. E. *Critical Stability Constants, Volume 2*. Plenum Press, New York, 1974.
24. Zhao, D., SenGupta, A. K. Ultimate removal of phosphate using a new class of anion exchangers. *Water Res.* **1998**, *32*, 1613–1625.
25. SenGupta, A. K., Zhu, Y., Hauze, D. Metal ion binding onto chelating exchangers with multiple nitrogen donor atoms. *Environ. Sci. Technol.* **1991**, *25*, 481–488.
26. Zhu, Y., Millan, E., SenGupta, A. K. Toward separation of toxic metal (II) cations by chelating polymers: Some noteworthy observations. *React. Polym.* **1990**, *13*, 241–253.
27. Zhao, D., SenGupta, A. K. Selective removal and recovery of phosphate in a novel fixed-bed process. *Water Sci. Technol.* **1996**, *33*, 139–147.
28. Pandit, A., Sengupta, S. Selective removal of phosphorus from wastewater combined with its recovery as a solid-phase fertilizer. *Water Res.* **2011**, *45*, 318–330.
29. Hoell, W. H., Feuerstein, W. Partial demineralization of water by ion exchange using carbon dioxide as regenerant Part III field tests for drinking water treatment. *React. Polym.* **1986**, *4*, 147–153.
30. Helfferich, F. G. Models and physical reality in ion-exchange kinetics. *React. Polym.* **1990**, *13*, 191–194.
31. Crank, J. *The Mathematics of Diffusion*. Oxford University Press, London, 1975.
32. Shin, D. H., Ko, Y. G., Choi, U. S., Kim, W. N. Design of high efficiency chelate fibers with an amine group to remove heavy metal ions and pH-related FT-IR analysis. *Ind. Eng. Chem. Res.* **2004**, *43*, 2060–2066.
33. Greenleaf, J. E., SenGupta, A. K. Environmentally benign hardness removal using ion exchange fibers and snowmelt. *Environ. Sci. Technol.* **2006**, *40*, 370–376.
34. Greenleaf, J. E., Lin, J., SenGupta, A. K. Two novel applications of ion exchange fibers: Arsenic removal and chemical-free softening of hard water. *Environ. Prog.* **2006**, *25*, 300–311.
35. Deng, S., Bai, R. B. Aminated poly(acrylonitrile) fibers for humic acid adsorption: Behaviors and mechanisms. *Environ. Sci. Technol.* **2003**, *37*, 5799–5805.
36. Matsumoto, H., Wakamatsu, Y., Minagawa, M., Tanioka, A. Preparation of ion-exchange fiber fabrics by electrospray deposition. *J. Colloid. Interface Sci.* **2006**, *293*, 143–150.
37. Soldatov, V. S., Shumkevich, A. A, Sergeev, G. I. Synthesis, structure and properties of new fibrous exchangers. *React. Polym. Ion Exch. Sorbents* **1988**, *7*, 159–172.

38. Dominguez, L., Benak, K. R., Economy, J. Design of high efficiency polymeric cation exchange fibers. *Polym. Adv. Technol.* **2001**, *12*, 197–205.

39. Deng, S., Yu, G., Xie, S., Yu, Q., Huang, J., Kuwaki, Y., Iseki, M. Enhanced adsorption of arsenate on the aminated fibers: Sorption behavior and uptake mechanism. *Langmuir* **2008**, *24*, 10961.

40. Jaskari, T., Vuorio, M., Kontturi, K., Manzanares, J. A., Hirvonen, J. Ion-exchange fibers and drug delivery: An equilibrium study. *J. Control. Release* **2001**, *70*, 219–229.

41. Yang, M., Tong, J. Loose ultrafiltration of proteins using hydrolyzed polyacrylonitrile hollow fiber. *J. Membr. Sci.* **1997**, *132*, 63–71.

42. Gupta, M. L., Gupta, B., Oppermann, W., Hardtmann, G. Surface modification of poly(acrylonitrile) staple fibers via alkaline hydrolysis for superabsorbent applications. *J. Appl. Polym. Sci.* **2004**, *91*, 3127–3133.

43. Zhang, G., Song, X., Li, J., Ji, S., Liu, Z. Single-side hydrolysis of hollow fiber polyacrylonitrile membrane by an interfacial hydrolysis of a solvent-impregnated membrane. *J. Membr. Sci.* **2010**, *350*, 211–216.

44. Bryjak, M., Hodge, H., Dach, B. Modification of porous polyacrylonitrile membrane. *Angew. Makromol. Chem.* **1998**, *260*, 25–29.

45. Deng, S., Bai, R. B. Adsorption and desorption of humic acid on aminated polyacrylonitrile fibers. *J. Colloid Interface Sci.* **2004**, *280*, 36–43.

46. Lohokare, H. R., Kumbharkar, S. C., Bhole, Y. S., Kharul, U. K. Surface modification of polyacrylonitrile based ultrafiltration membrane. *J. Appl. Polym. Sci.* **2006**, *101*, 4378–4385.

47. Zhang, G., Meng, H., Ji, S. Hydrolysis differences of polyacrylonitrile support membrane and its influences on polyacrylonitrile-based membrane performance. *Desalination* **2009**, *242*, 313–324.

48. Wang, J., Yue, Z., Economy, J. Solvent resistant hydrolyzed polyacrylonitrile membranes. *Sep. Sci. Technol.* **2009**, *44*, 2827–2839.

49. Shunkevich, A. A., Akulich, Z. I., Mediak, G. V., Soldatov, V.S. Acid-base properties of ion exchangers. III. Anion exchangers on the basis of polyacrylonitrile fiber. *React. Funct. Polym.* **2005**, *63*, 27–34.

50. Zhang, S., Chen, S., Zhang, Q., Li, P., Yuan, C. Preparation and characterization of an ion exchanger based on semi-carbonized polyacrylonitrile fiber. *React. Funct. Polym.* **2008**, *68*, 891–898.

51. Reneker, D. H., Yarin, A. L., Fong, H., Koombhongse, S. Bending instability of electrically charged liquid jets of polymer solutions in electrospinning. *J. Appl. Phys.* **2000**, *87*, 4531–4547.

52. Gibson, P. W., Schreuder-Gibson, H. L., Rivin, D. Transport properties of porous membranes based on electrospun nanofibers. *Colloid Surface A* **2001**, *187–188*, 469–481.

53. Dzenis, Y. Spinning continuous fibers for nanotechnology. *Science* **2004**, *304*, 1917–1919.

54. Ramakrishna, S., Fujihara, K., Teo, W. E., Yong, T., Ma, Z., Ramaseshan, R. Electrospun nanofibers: Solving global issues. *Mater. Today* **2006**, *9*, 40–50.

55. Kaur, S., Kotaki, M., Ma, Z., Gopal, R., Ramakrishna, S., Ng, S. C. Oligosaccharide functionalized nanofibrous membrane. *Int. J. Nanosci.* **2006**, *5*, 1–11.

56. Huang, Z., Zhang, Y., Kotaki, M., Ramakrishna, S. A review on polymer nanofibers by electrospinning and their applications in nanocomposites. *Compos. Sci. Technol.* **2003**, *63*, 2223–2253.

57. Teo, W. E., Ramakrishna, S. A review on electrospinning design and nanofiber assemblies. *Nanotechnology* **2006**, *17*, R89–R106.

58. An, H., Shin, C., Chase, G. G. Ion exchanger using electrospun polystyrene nanofibers. *J. Membr. Sci.* **2006**, *283*, 84–87.

59. Neghlani, P. K., Rafizadeh, M., Taromi, F. A. Preparation of aminated-polyacrylonitrile nanofiber membranes for the adsorption of metal ions: Comparison with microfibers. *J. Hazard. Mater.* **2011**, *186*, 182–189.

60. Liu, H., Hsieh, Y. Preparation of water-absorbing polyacrylonitrile nanofibrous membrane. *Macromol. Rapid. Commun.* **2006**, *27*, 142–145.

61. Sutasinpromprae, J., Jitjaicham, S., Nithitanakul, M., Meechaisue, C., Supaphol, P. Preparation and characterization of ultrafine electrospun polyacrylonitrile fibers and their subsequent pyrolysis to carbon fibers. *Polym. Int.* **2006**, *55*, 825–833.

62. Wang, T., Kumar, S. Electrospinning of polyacrylonitrile nanofibers. *J. Appl. Polym. Sci.* **2006**, *102*, 1023–1029.

63. Moroni, L., Licht, R., de Boer, J., de Wijn, J. R., van Blitterswijk, C. A. Fiber diameter and texture of electrospun PEOT/PBT scaffolds influence human mesenchymal stem cell proliferation and morphology, and the release of incorporated compounds. *Biomaterials* **2006**, *27*, 4911–4922.

64. Chiu, H. T., Lin, J. M., Cheng, T. H., Chou, S. Y. Fabrication of electrospun polyacrylonitrile ion-exchange membranes for application in lysozyme. *eXPRESS Polym. Lett.* **2011**, *5*, 308–317.

65. Chen, M., Patra, P. K., Warner, S. B., Bhowmick, S. Optimization of electrospinning process parameters for tissue engineering scaffolds. *Biophys. Rev. Lett.* **2006**, *1*, 153–178.

66. Bajaj, P., Kumari, M. S. Structural investigations on hydrolyzed acrylonitrile terpolymers. *Eur. Polym. J.* **1988**, *24*, 275–279.

67. Litmanovich, A. D., Plate, N. A. Alkaline hydrolysis of polyacrylonitrile. On the reaction mechanism. *Macromol. Chem. Phys.* **2000**. *201*, 2176–2180.

68. Wang, Z., Wan, L., Xu, Z. Surface engineering of polyacrylonitrile-based asymmetric membranes towards biomedical applications: An overview. *J. Membr. Sci.* **2007**, *304*, 8–23.

69. Oh, N., Jegal, J., Lee, K. Preparation and characterization of nanofiltration composite membranes using poly(acrylonitrile) (PAN). I. Preparation and modification of PAN supports. *J. Appl. Polym. Sci.* **2001**, *80*, 1854–1862.

70. Oh, N., Jegal, J., Lee, K. Preparation and characterization of nanofiltration composite membranes using poly(acrylonitrile) (PAN). II. Preparation and characterization of polyamide composite membranes. *J. Appl. Polym. Sci.* **2001**, *80*, 2729–2736.

71. Young, R. A. Cross-linked cellulose and cellulose derivatives. In *Absorbent Technology*, Chatterjee, P. K., Gupta, B. S., Eds. Elsevier Science, Amsterdam, the Netherlands, 2002.

72. Morel, F. M. M., Hering, J. G., *Principles and Applications of Aquatic Chemistry*, Wiley-Interscience, New York, 1993.

73. Sengupta, S., SenGupta, A. K. Trace heavy metal separation by chelating ion exchangers. In *Environmental Separation of Heavy Metals: Engineering Processes*. SenGupta, A. K., Ed., CRC Press, Boca Raton, FL, 2002.

74. Hudson, M. Coordination chemistry of selective ion-exchange resins. In *Ion Exchange: Science and Technology*. Rodrigues, A. E., Ed., NATO ASI Series, Martinus Nijhoff, Boston, 1986.

75. Warshawsky, A. Modern research in ion exchange. In *Ion Exchange: Science and Technology*. Rodrigues, A. E., Ed., NATO ASI Series, Martinus Nijhoff, Boston, 1988.

76. Vuorio, M., Manzanares, J. A., Murtomaki, L., Hiroven, J., Kankkunen, T., Kontturi, K. Ion-exchange fibers and drugs: A transient study. *J Control. Release* **2003**, *91*, 439–448.

77. Stevens, T. S., Davis, J. C. Hollow fiber ion-exchange suppressor for ion-chromatography. *Anal. Chem.* **1981**, *53*, 1488–1492.

78. Chen, L., Yang, G., Zhang, J. A study on the exchange kinetics of ion-exchange fiber. *React. Funct. Polym.* **1996**, *29*, 139–144.

5 Separation of Concentrated Ion Mixtures in Sorption Columns with Two Liquid Phases

Ruslan Kh. Khamizov, Anna N. Krachak,
Alex N. Gruzdeva, Sultan Kh. Khamizov,
and Natalya S. Vlasovskikh

CONTENTS

A new approach to the treatment of highly concentrated solutions of electrolytes in ion exchange and sorption columns with zero bed porosity is proposed. The approach concerns reducing the inter-particle space occupied by working solutions in columns by filling them with special organic liquids immiscible with water.

By a variety of examples, including the separation of nitric acid and nitrates for processing aluminum concentrates, purification of working electrolytes for copper refining, conversion of potassium chloride into chlorine-free fertilizers, and so on, the advantages of the proposed method are demonstrated and show that using the new method in ion exchange and acid retardation processes does not affect the equilibrium separation parameters, but significantly improves the kinetics by eliminating longitudinal dispersion effects. This also makes it possible to reduce the duration of working cycles, thereby raising the efficiency of the use of separation columns.

5.1 INTRODUCTION

For many years after the beginning of the industrial use of ion exchangers, an opinion existed that these materials are intended only for the separation and concentration of components contained in solutions at relatively low concentrations. The most typical examples of such uses of ion exchangers are water treatment for power plants, the removal of microcomponents in hydrometallurgy, and various methods for solution purification. As the rule, these technologies are characterized by prolonged filtration cycles, that is, the volumes of solutions processed in one sorption cycle are many times larger than the volume of a sorbent bed.

Today, along with such technologies, the treatment of concentrated electrolyte solutions with ion exchangers gains more and more practical importance. Such processes have been proposed at different times, but all are characterized by short filtration cycles, for which the volumes of processed solutions and sorbent bed are comparable. There are a number of examples of such processes; three are presented in the following.

In the early 1960s, one of the first publications[1] concerned the possibility of separating acids from their salts in highly concentrated solutions on ion exchangers. The method, called "acid retardation" (AR) by its authors, was not an ion exchange, because the isolation of the hydronium compound from the derivatives of metal ions was not performed by a cation exchanger; but an anionic resin in a ion form similar to the common anion of separated components was applied. The method proved to be attractive due to the fact that desorption was carried out with water. For a long time, this method did not find industrial application due to difficulties in organizing the flow distribution in columns to eliminate mixing of the different substances. Since the late 1970s, after process modifications, it began to be applied and is now widely used, for example, by the Canadian company Eco-Tec.[2,3] Research in AR has expanded the range of separable components[4] and the development of new sorption materials.[5]

Another example is the ion exchange conversion of chloride-containing fertilizers into chlorine-free products, as well as the production of complex NPK fertilizers, using highly concentrated solutions. Scientific research has been actively conducted since the beginning of the 1970s[6] and today one can find examples of the industrial use of different variants of ion exchange technology.[7]

We also believe that using the effect of isothermal supersaturation of solutions (IXISS effect) for inorganic electrolytes[8–10] is interesting for the practice of ion exchange technologies. The essence of this effect lies in the fact that with a dynamic ion exchange process accompanied by the formation of a slightly soluble substance, a relatively stable supersaturated solution is generated in the sorbent bed, and the spontaneous crystallization of this substance takes place in the effluent. A lot of processes in which the ion exchange reaction product is less soluble than the initial compound can be found. Among them, one can also find a lot of practically feasible processes in which the temporary stabilization of a supersaturated solution in a packed bed of ion exchanger continues for a few hours or more.[9] In the processing of highly concentrated solutions, the opportunity to obtain directly a final high-purity product in a solid state without additional energy consumption is very attractive. In practice, this approach proved to be indispensable in the technologies of zero-discharge processing of sea-water and sea brines.[9]

As can be seen from these examples, separation of concentrated ionic mixtures in aqueous solutions can be carried out by ion exchange or non-exchange processes on resins. For these and other processes with small cycle times for the sorption and regeneration stages, as well as for processes where the solutions in these stages differ in density or temperature, we encounter a main problem: impairment of separation due to the influence of mixing effects caused by longitudinal dispersion in columns (all types of additional erosion of concentration profiles other than molecular diffusion).[11,12] Well-known approaches can be applied to eliminate dispersion effects, for example, the more concentrated solution is fed to the column in the upward direction; a hot solution is fed from top to bottom, special equipment and conditions are chosen to reduce the inter-particle space and free volume under the bed. At the same time, an effect that is hard to avoid is bound up with the shrinking and swelling of resin beads. This phenomenon is illustrated in Figure 5.1a; the previous contact of a lower bead in the column with a more dense solution results in its shrinking, leading to the channeling and subsequent shrinking of the upper bead, and so on. One of the ways to eliminate such a progressive erosion of the concentration frontiers and to improve flow distribution in resin beds is proposed by Craig Brown.[2] This is achieved by over-packing a special column with highly compressed and deformed resin beads to reduce the inter-particle space (porosity volume) of the ion exchanging bed shown by the upper arrow in Figure 5.1b.

A new approach to the organization of mass transfer processes in ion exchange and sorption beds is described in this chapter. The approach consists in reducing the space occupied by water and working solutions in the columns by using an additional liquid phase.[13] The column, loaded with ion exchange granulated material, is additionally filled by an organic liquid immiscible with the aqueous solutions. No special facilities and no high pressures are required and the ion exchange materials can be used for long time. All the column processes can be carried out in down-flow mode for both sorption and regeneration stages of the working cycle.

The proposed approach offers another opportunity that is essential for the processing of complicated technological solutions: in the course of such a processing, supersaturated solutions and colloidal systems can be formed, which ultimately leads to the formation of suspensions. In this case, the application of over-packed beds is almost impossible. At the same time, as will be seen from the results presented here,

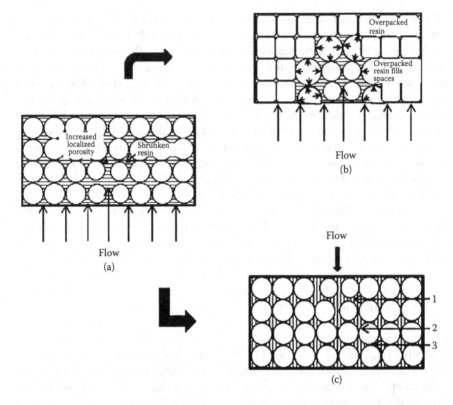

FIGURE 5.1 Two ways of reducing longitudinal dispersion in columns: (a) system with progressive erosion of the concentration profile; (b) over-packed bed; (c) bed filled with liquid organic phase: 1, organic fluid; 2, working solution; 3, contact point. [(a) and (b) are taken from Brown, C. J., Fluid treatment method and apparatus, US Patent No 4673507.]

the use of sorption columns with two immiscible liquid phases allows solving many complex tasks.

In this chapter, we demonstrate the prospects of the proposed approach by a variety of actual examples for chemical engineering and elemental analysis.

5.2 PERCOLATION THROUGH A SORPTION BED WITHOUT INTER-PARTICLE SPACE

At first glance, the column seems to be unusable if the granular bed of ion exchanger is completely filled with organic liquid, and no free volume remains in it including the inter-particle space. However, experimental tests show such a possibility, when using a low-dense organic liquid immiscible with water, for example, decanol or dodecane, and a hydrophilic sorption material, such as a strong-basic anion exchanger. In such a system, spontaneous water or aqueous solution passing through the column can be organized. One can see in this case that at the beginning of the process, a small amount of the organic liquid (a few percent of the bed volume) goes from the column and then only the aqueous effluent flows. Organic liquid remains in the column. The diagrams shown in Figure 5.2 are helpful to understand the process. As shown in Figure 5.2a, in a wide vessel with oil, a denser drop of water or aqueous solution falls, displacing the organics up. By decreasing the diameter of the vessel until it becomes a closed bottom capillary (Figure 5.2b), with the proviso that water has a higher energy of wetting the vessel walls than oil, the drop deforms and flows down on the wall surface, displacing the oil. Now take an open capillary of complicated shape shown in Figure 5.2c, along the axis of which there are periodic constrictions, forming a very thin slits. At a certain very small thickness of such a slit, water or the aqueous solution can occupy (from energy consideration) and go in it, and it is possible to organize the process of spontaneous flow through the capillary with the retention of oil therein. Percolation through a fluidized bed of granular ion exchanger may be similar to that shown in the process of a wide vessel. A plurality of capillaries like that one shown in Figure 5.2c, arranged in parallel, simulate a stationary resin bed, and it can be assumed that the passage of aqueous fluids through the bed takes on such a mechanism.

Thus, the following interpretation of mass transfer mechanisms in the proposed system can be given. On passing ion-containing solutions through a bed of granular ion exchange material in the absence of free space between the grains, as well as under conditions where the organic liquid repels water or the aqueous solution, the latter, being introduced at the interface between the organic liquid and the hydrophilic surface of the granules, forms very thin films enveloping each ion

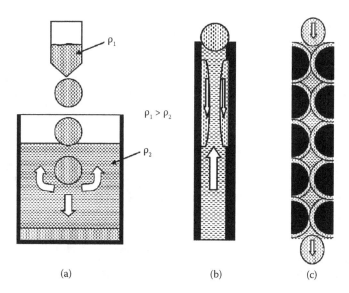

(a)   (b)   (c)

FIGURE 5.2 The flow of dense fluid through more light liquid: (a) in the bulk volume, (b) in a limited volume (through the capillary material with a hydrophilic surface), and (c) through the capillary modeling with a sorption bed.

exchange bead and drains by the points of contact between them (Figure 5.3b). Such a steady-state dynamic system of ultra-thin mobile films provides an extremely well-developed surface area of contact between the phases taking part in the mass exchange process.

Owing to the fact that working solutions occupy negligible space volume in the column, there is no possibility of the flows of different substances mixing, and also no problems arise with the choice of direction of these flows. The simplest variant is using a co-current downward regime.

At low flow rates, the organic liquid remains inside the column, but for hydrodynamic reasons, such a loss can also take place on operation with too high flow rates. However, the lighter organic liquid, on leaving the column, seeks immediately to go up and can be returned to the column by a tube connecting the bottom and top. For relatively large columns, a more complex system shown in Figure 5.3a can be applied. In this case, the bottom and the top of the column are connected by a special tube, which is provided with an apparatus for the separation of organic liquids from aqueous solutions having an input for shared emulsion to be separated out for the organic liquid substance and an outlet for the aqueous solution purified of the organic liquid substance. It uses in line of this connection a special minipump (mounted above said apparatus) to return organic liquid to the top of the column.[13-15]

Some examples are presented here, and they illustrate the possibilities of the proposed method, which we called the "NewChem method," from the name of the company, which played an important role in its development.[13-18]

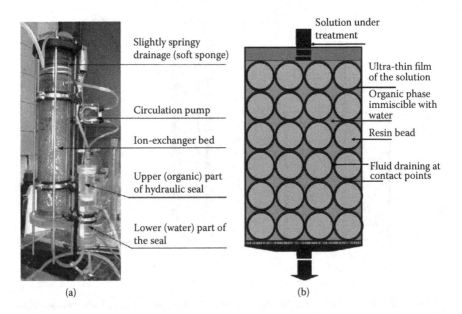

FIGURE 5.3 NewChem system for separation of acids and salts. (a) Photograph of a column (nanoporous reactor) with organic phase. (b) Sectional diagram of the column loading.

5.3 SEPARATION PROCESSES OF AR TYPE

Despite the fact that the AR process has been known for a long time, this theory is not developed enough, and there is still no consensus on the mechanism of separation. The same can be said about the NewChem method, because the presence of additional inert liquid phase in sorption bed only affects the hydrodynamics and mass transfer kinetics, but has no influence on equilibrium and selectivity.

Some authors[19] consider the separation by AR as a consequence of the exclusion of hydrated ions with relatively large size in nano-porous media. We undertook an attempt to develop the theory of AR processes[20,21] for ion exchanging materials (taking into account that gel-type resins can also be considered as the nanoporous media) and assume that the mechanism of separation is more complex, and additionally to the exclusion effect, includes sorption of acid molecules, as well as the competitive solvation of ion pairs in the resin phase. Here, we do not give the details of the proposed method[21] for calculating the dielectric constant and the degree of dissociation of the electrolytes in the sorbent phase and on this basis, on a mathematical model describing the existing experimental results. As to the exclusion effect on gel-type resins, we stay on neutral molecules and ion pairs. In the mixed electrolytes, concentrated acids form bonded ion pairs or molecules, which are poorly hydrated. Therefore, the molecular and ionic species of acids are sufficiently small to penetrate into the nanopores, that is, the acid may "impregnate" nanoporous granules in the reactor, the column with an equilibrated ion exchanger. The dissolved salts remain more dissociated even at conditions of reduced dielectric permittivity and form larger highly hydrated ions or slightly associated pairs, which cannot penetrate into the pores and pass without delay through the resin bed. The last one can be considered as the nano-porous reactor, which competitively retards molecules of water and the hydrated molecular and ionic particles of acid and salts.

5.3.1 PROCESSING OF LIQUID CONCENTRATES OF ALUMINUM LEACHING

Consider the effect of using the NewChem method for some practical processes. The first one concerns the processing of the nepheline concentrate for producing aluminum and mineral fertilizers by treating solutions produced by acid leaching. The problem here is the need to separate excess acid from these solutions to return it to the head of the process, thus leading to significant savings in acid, alkali, and water. It would also facilitate the purification of the aluminum solution from iron that is easier to be carried out in weakly acidic than in concentrated acid solutions.

Figure 5.4 shows two different concentration histories for the separation of nitric acid and aluminum nitrate in the nitrate form with a strong acid anion exchanger. The upper picture shows the results obtained using the NewChem approach while the second represents experimental results obtained with a standard column, without an organic phase, and it is given for comparison. Complete working cycles of the AR method are also shown in this figure.

At the sorption stage, the actual liquid concentrate produced by acid leaching of nepheline ore is used as the feed solution, and regeneration (acid desorption) is performed by DI water. Different stages are separated in the figure by vertical dashed lines.

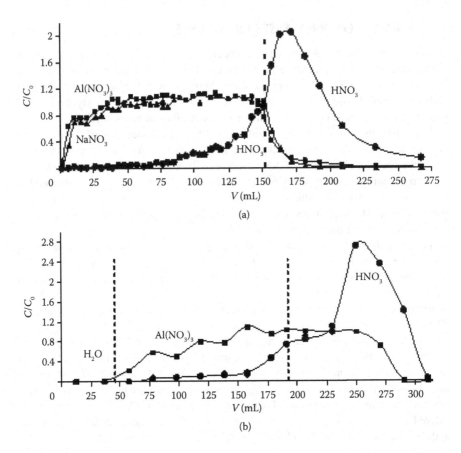

FIGURE 5.4 Breakthrough concentration curves of components in the separation of nitric acid and metal nitrates. Column loadings: 110 mL strong base anionic resin AV-17 NO_3 form. Composition of the influent at sorption stage (mass %): H_2O, 67; $Al(NO_3)_3$, 16; HNO_3, 7.3; $NaNO_3$, 5.9; KNO_3, 2.1; $Fe(NO_3)_3$, 1.2; desorbing agent, water; flow rate, 2 BV/h. (a) NewChem technique; (b) standard column.

The top picture illustrates that only the stages of sorption and regeneration take place, and the lower one shows the additional initial stage of water leaving the column bed.

With the other process conditions being equal, the new approach not only leads to a reduced duration of the working cycle but also increases the efficiency of separation.[13–15]

Analysis of experimental results shows that the use of the NewChem method for AR and separation process does not affect the equilibrium parameters of separation, but significantly influence the kinetics by the elimination of longitudinal dispersion effects.[16,18]

Another interesting properties of the process is shown in Figure 5.5, which also shows concentration histories for the separation of nitric acid and aluminum nitrate in the nitrate form with the anionic resin. At a sorption stage, the actual

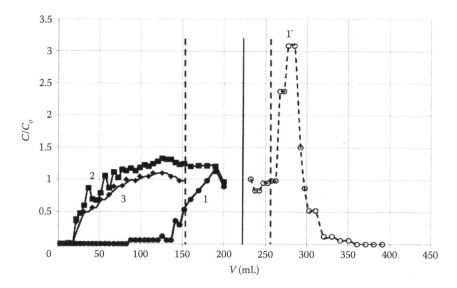

FIGURE 5.5 Breakthrough concentration curves of HNO_3 (1, 1′) and dissolved aluminum nitrate (2) in one of the repetitive separation cycles of the acid process. (3) Concentration of Fe in effluent samples before the precipitation of iron hydroxide. Laboratory column loadings: 110 mL of granulated strong base anionic resin AV-17 of gel type in NO_3 form. Organic phase: decanol. Flow rate: 2 BV/h.

liquid concentrate produced by acid leaching of nepheline concentrate was used as the feed solution; the acid desorption stage was performed by DI water. Effluent volumes corresponding to these different stages are separated by vertical solid straight line in the figure. The dashed vertical lines indicate the left and right bounds within which the effluent fractions are to be returned into the separating reactor. Owing to the deep separation at AR, the effluent is found to be neutral or weakly acidic (pH > 2 up to 130 mL). It leads to the formation of a colloidal solution while passing through the bed and to spontaneous precipitation of iron hydroxide after leaving the column. Checking the balance of separated acid shows its amount exceeding the content of the free acid in raw solution. It takes place owing to the effect of "soft hydrolysis," which can take place as follows:

$$[R–NO_3 \cdots H_2O] + HNO_3 = [R–NO_3 \cdots HNO_3] + H_2O \tag{5.1}$$

$$3[R–NO_3 \cdots H_2O] + Fe(NO_3)_3 = Fe(OH)_3\downarrow + 3[R–NO_3 \cdots HNO_3] \tag{5.2}$$

$$[R–NO_3 \cdots HNO_3] + H_2O = [R–NO_3 \cdots H_2O] + HNO_3 \tag{5.3}$$

where R denotes the resin group that can be associated with water or acid molecule. Processes in Equations 5.1 and 5.2 take place at the sorption stage, while the process in Equation 5.3 at acid desorption.

The distinction of real technological process, consisting of repetitive cycles illustrated in Figure 5.5, is that the effluent fractions of the final part of the sorption stage and the starting part of desorption stage (fraction volumes between the dashed vertical lines) are to be combined and recycled to the head of the process. In addition to this procedure, the stage of acid desorption with water is to be stopped at any point of effluent concentration that is convenient for the technological aims.

Thus, when using acidic digestion technology, the NewChem method allows returning the residual acid in the process, carrying out additional purification of work solution from iron, carrying out (partially) low-energy consuming process of soft hydrolysis (instead of thermo-hydrolysis), and additionally returning the corresponding amount of acid in the process.[22]

Another example of a topical technological task is the processing of circulating solution at acid-salt leaching of aluminum from low-grade bauxites. Standard industrial alkaline methods for alumina production (Bayer process and its modifications) require low-silica raw materials which become inaccessible due to the gradual depletion of high-grade bauxite deposits.[23] An ammonium hydrosulfate (bisulfate) process called "Aloton" or "Buchner"[24] is considered today as one of the most prospective methods. The technique consists in implementing a circular process comprising the steps of thermal decomposition of ammonium sulfate into ammonia and ammonium bisulfate, dissolving the latter to treat the aluminum-containing feedstock, filtering the solution of alum produced, cooling the filtrate and separating alum from the circulating solution, dissolving the alum in pure water and precipitating aluminum hydroxide with ammonia, and isolating ammonium sulfate from the mother liquor to return it to the head of circuit. The problem is in the removal of acid from the recirculating solution and the partial purification of it from iron.

The results of the NewChem treatment are shown in Figure 5.6. The process was performed with the use of a pilot-scale column with a bed volume of 30 L. The following processes take place; Equations 5.4 and 5.5 occur at sorption stage and Equation 5.6 at acid desorption:

$$[R\text{–}SO_4 \cdots H_2O] + 2NH_4HSO_4 = (NH_4)_2SO_4 + [R\text{–}SO_4 \cdots H_2SO_4] \qquad (5.4)$$

$$[R\text{–}SO_4 \cdots H_2O] + FeSO_4 + H_2O + O_2 = Fe(OH)_3\downarrow + [R\text{–}SO_4 \cdots H_2SO_4] \quad (5.5)$$

$$[R\text{–}SO_4 \cdots H_2SO_4] + H_2O = [R\text{–}SO_4 \cdots H_2O] + H_2SO_4 \qquad (5.6)$$

The view of concentration curves is similar to that for nitric AR described previously. Experimental conditions are also the same, excluding that the feed solution is to be aerated at the inlet of the column. In parallel to sulfuric AR on the sulfate form of the anion exchanger, an additional process of "soft" hydrolysis takes place, which partially removes iron from the work solution and enhances acid recirculation in the process.

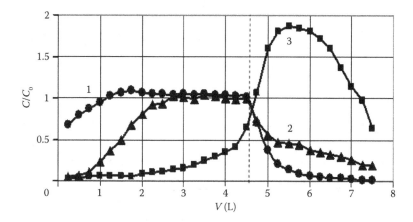

FIGURE 5.6 Breakthrough curves of some components in the process of arsenic removal from copper electrolyte. 1, $CuSO_4$; 2, H_2SO_4; 3, $H_3AsO_3 + HAsO_3$. [Column loadings: 3000 mL AV-17 strong base anionic resin in initial form equilibrated with pure electrolyte and washed after with water. Composition of the influent at sorption stage (g/L): H_2SO_4, 220; $CuSO_4$, 105; total As, 4.5 mg/L. Desorbing agent: water. Flow rate: 2 BV/h.]

5.3.2 PURIFICATION OF WORKING ELECTROLYTES FOR COPPER REFINING

The following example concerns a working electrolyte in the refining of copper by electrolysis. During the operation, the electrolyte is polluted by various impurity components including arsenic, which seriously affects the quality of the product. Figure 5.7 shows the concentration curves of one of the steady-state cycles of pilot tests carried out with real electrolyte from the electrolysis section of the Balkhash copper plant (Kazakhstan). The second column in Table 5.1 shows the main composition of work electrolyte to be treated.

The tests were conducted in the apparatus shown in Figure 5.3. Preliminarily, the anion exchanger was equilibrated with work electrolyte to obtain a mixed anionic form (sulfate with a minor part of other anionic forms) and then, the resin was washed with water to use it in repetitive working cycles and to avoid the ion exchange processes during them.

The major AR processes for a sorption stage can be illustrated as follows:

$$\left[R\text{--}SO_4 \ (R\text{--}HAsO_4) \cdots H_2O \right] + H_2SO_4 \ (H_3AsO_4)$$
$$= H_2O + \left[R\text{--}SO_4 \ (R\text{--}HAsO_4) \cdots H_2SO_4 \ (H_3AsO_4) \right]$$

(5.7)

To perform the separation process, the effect of competitive sorption of sulfuric acid and the As(III) and As(V) acids is used here.

As seen in Figure 5.7, arsenic derivatives are retarded much more selectively. The As concentrate produced at the desorption stage did not contain too much of the target component, copper, and this solution can be recycled into a further operational process for arsenic utilization. The pilot test, which was carried out with the use of

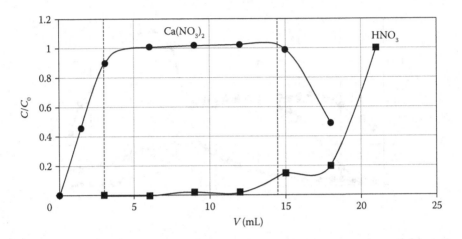

FIGURE 5.7 Concentration curves for the separation of nitric acid and metal nitrates. Analytical column bed: $L = 25$ cm, $S = 1$ cm^2 of strong base anionic resin AV-17 in NO_3 form. Composition of the initial model solution (mol/L): HNO_3, 5.89; $Ca(NO_3)_2$, 1.47; Fe (II), 1×10^{-3}; Mn, 1×10^{-3}; Cu, 1×10^{-4}; Zn, 2×10^{-4}; Co, 1×10^{-4}. Desorbing agent: water. Flow rate: 1.2 BV/h.

TABLE 5.1
Composition of Contaminated and Processed Work Electrolytes

Component	Initial Concentrations, g/L	Final Concentrations, g/L
Cu^{2+}	40.30	41.2
Ni^{2+}	2.25	0.75
SO_4^{2-}	204.40	209.50
As(III)+As(V)	4.41	0.86
Fe (II,III)	2.50	0.83
Ca^{2+}	0.54	0.36
$CuSO_4$	104	104
H_2SO_4	122	122

two-step processing with the use of the AR method, has led to the purification of the electrolyte to the concentrations shown in the third column of Table 5.1.

5.3.3 PURIFICATION OF INDUSTRIAL PHOSPHORIC ACID WITH SEPARATION OF REM

The main commercial technology for the production of phosphoric acid is the "wet process" based on the interaction of hot concentrated sulfuric acid with naturally occurring phosphate rocks.[25] Such an industrial acid is contaminated by a variety of macro- and microcomponents, and it is usually used for the production of ordinary (insoluble) phosphoric or complex fertilizers. Partial purification of wet phosphoric acid (WPA) improves its quality and extends the areas of application. For example, refined WPA can be used for the production of soluble fertilizers for drop fertigation and intense crop production. On the other hand, industrial WPA contains valuable components, such as rare earth metals (REMs), and searching for the combined processes of their separation with the acid purification is very important.[26,27]

Tables 5.2 and 5.3 show an example of the composition of industrial wet process phosphoric acid manufactured in one of the Russian plants, located in the city of Cherepovets. The admixture of macrocomponents, other than P_2O_5, makes more than 5% of the product mass. The total content of REMs is more than 0.1% (1 g/kg).

TABLE 5.2
Content of Selected Macrocomponents in the Wet Process Phosphoric Acid Manufactured at Ammophos, Russia

Component	Content, %
P_2O_5	26.6
SO_3	2
F	1.8
CaO	0.1
Na_2O	0.06
K_2O	0.01
SiO_2	1.1
Al_2O_3	0.44
Fe_2O_3	0.26
MgO	0.053

TABLE 5.3
Content of REM in the Wet Process Phosphoric Acid Manufactured at Ammophos, Russia

Element	Concentration, mg/kg (ppm)
Ce	311
Nd	158
La	128
Pr	42
Gd	28
Sm	27
Dy	16
Eu	8.3
Er	7.0
Yb	4.0
Tb	3.6
Ho	2.8
Tm	0.76
Lu	0.48

Figure 5.8a and b shows the process of separation of salt components and phosphoric acid. For performing such an experiment, first, a strong acid anionic resin in the column was treated with initial WPA until equilibrium was achieved, and then, the column was washed with diluted water solution of phosphoric acid. It is interesting to note that some of salt components like the salts of calcium, magnesium,

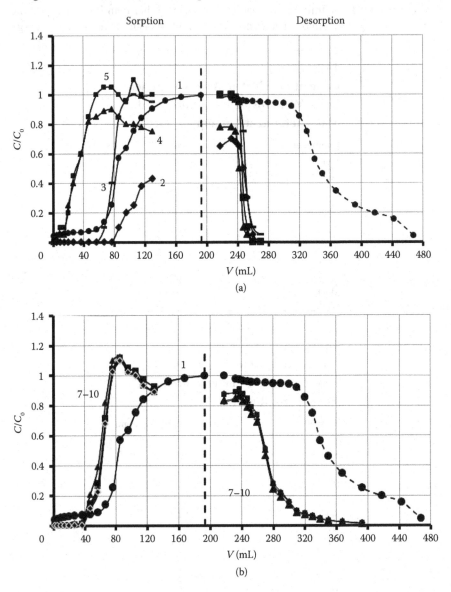

FIGURE 5.8 Separation of the components of industrial phosphoric acid by acid retardation frontal chromatography. (a) 1, acid; 2: titanium; 3, iron; 4, manganese; 5, calcium and magnesium. (b) 7–10: yttrium, lanthanum, cerium, neodymium column with a bed of AB-17 (174 mL) in an equilibrated mixed form. Organic phase: dodecane.

and manganese leave the column faster than the compounds of iron and REMs (Figure 5.8b). It can be proposed that this effect is related to the balance between cationic salts and anionic complexes, which can be formed by the last elements. One can also see that deeply refined phosphoric acid is produced at the stage of its desorption with diluted water solution. Figure 5.8 illustrates the process, in which the stages of sorption and desorption are continued until equilibrium is reached. It is made for visualization of the whole process. In a real technological process, the effluent fractions corresponding to the ending phase of sorption and the starting phase of the desorption stage are combined, recycled, and added to the initial WPA solution. The middle portions of desorbed acid are used as the purified product, and the final portions of the effluent, being more diluted but more pure phosphoric acid, are used for additional dilution and preparation of the displacing solution to apply it (instead of water) for the desorption of acid in the next working cycle. As to the separation of REMs (at the sorption stage), the portions of effluent between the points of acid concentrations: $0.15 < C/C_0 < 0.65$ are collected, treated with ammonia to adjust to pH 2.5, and filtrated to obtain initial (crude) concentrate to recover REM and filtrate to be used for fertilizer production.[28]

Two series of pilot experiments have been carried out with the use of columns loaded with 10 L of strong base anion exchanger and filled with dodecane. Some of the data obtained are presented in Tables 5.4 and 5.5, and these results demonstrate

TABLE 5.4

Composition of Primary REM—Concentrate Produced at Pilot Tests with WPA from Ammophos

Oxide	mg/kg (ppm)	%
Al_2O_3	35,053.33	3.51
CaO	45,787.50	4.58
Ce_2O_3	156,200.00	15.62
Fe_2O_3	84,425.00	8.44
La_2O_3	52,855.00	5.29
MgO	21,175.00	2.12
MnO	14,877.50	1.49
Na_2O	124,850.00	12.49
Nd_2O_3	84,700.00	8.47
SrO	8635.00	0.86
Other oxides of REM	65,045.20	6.50
TiO_2	77,000.00	7.70
Y_2O_3	29,865.00	2.99

Note: The effluent of the NewChem Process between the points of acid concentrations $0.15 < C/C_0 < 0.65$ has been collected, treated with ammonia to adjust to pH 2.5, and filtrated to obtain initial (crude) concentrate, which has been dried.

TABLE 5.5

Content of Selected Macrocomponents in Initial and Purified WPA

Macrocomponent	Initial WPA	WPA Purified in Successive and Repetitive Cycles					
		Cycle 1	Cycle 2	Cycle 3	Cycle 4	Cycle 5	Cycle 6
P_2O_5, %	25.39	22.84	23.02	22.49	22.01	22.67	23.02
CaO, %	0.47	0.055	0.07	0.11	0.08	0.07	0.05
MgO, %	1.21	0.088	0.09	0.12	0.07	0.06	0.08
SO_3, %	1.4	1.24	1.17	1.05	1.12	1.07	1.09

Note: The data are taken from the protocol of joint pilot tests carried out at EuroChem-BMU, Russia.

good prospects for the development of technology to process industrial phosphoric acid using the proposed variant of the AR method. Preliminary estimations show that two large-scale columns, each 10 m³, working in opposite phases to ensure continuous operation, can provide more than 50,000 tons per year of purified phosphoric acid and tens of tons of REM concentrate (dependently of their initial content in raw acid).

5.3.4 PROCESSING OF REM- AND Zr-CONTAINING NITRATE CONCENTRATE PRODUCED AT EUDIALYTE DECOMPOSITION

In contrast to phosphoric acid media, in mixed concentrated solutions of nitric acid and nitrates of metals, the difference between the elements forming and not forming strong anionic complexes becomes significantly more noticeable, and separation of the compounds of REM from nitric acid is much more difficult.[29] This effect can be used to arrange other types of separating valuable components in mixed solutions, for example, Zr and REM, which are leached together in acidizing eudialyte ore.[30] The example of liquid concentrate composition, which can be obtained after the digestion of eudialyte with concentrated nitric acid, is given in Table 5.6.

Figures 5.9 and 5.10 show examples of concentration curves of sorption and desorption of components from the studied nitrate solution in the separation system with two liquid phases. Experiments have been performed in a column with 150 mL of commercial anion exchanger AV-17 in nitrate form filled with dodecane used as the organic phase, at a flow rate of 1 BV/h for all the stages.

The majority of macrocomponents of metal salt type pass through the column with a flow rate near to that of pumping the sample solution; they leave the column being composed of a neutral or slightly acidic solution. Nitric acid is retarded, and its concentration begins to rise only after the majority of metal concentrations achieve equilibrium enrichment (the corresponding concentrations in the initial work solution). At the desorption stage, performed by washing the column with deionized water, nitric acid of varying concentrations is removed, and the peak of

TABLE 5.6

Content of Selected Nitrates and Nitric Acid in the Liquid Concentrate of Eudialyte Digestion, g/L

Al	Ca	Ce	Fe	La	Sr	Y	Zr	HNO_3
15.84	50.34	2.04	101.19	0.61	12.76	2.31	35.94	88.20

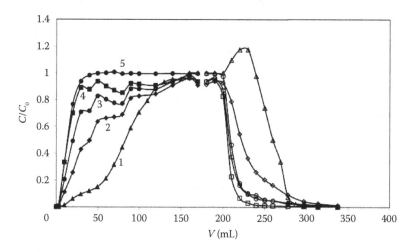

FIGURE 5.9 Concentration histories for HNO_3 (1), Ca (2), Fe (3), Zr (4), and Al (5) during separation of the components of nitric acid concentrate of eudialyte digestion. The same points hold good for sorption (with numbers) and desorption stages.

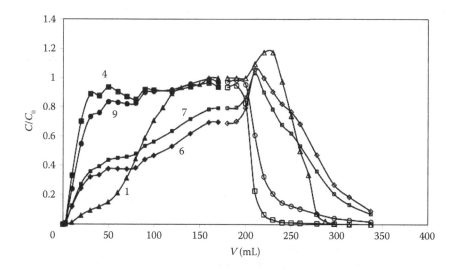

FIGURE 5.10 Concentration histories for HNO_3 (1), Zr (4), La (6), Ce (7), and Y (9) for the experiment shown in Figure 5.9.

the desorption curve is slightly higher than its content in the work solution. A sharp decrease in the concentration of salts takes place at this stage. Test showed the results of parallel experiments having good repeatability.

Interestingly, the rare elements behave differently in both stages of the process. The behavior of Y and Zr salts is analogous to other macrocomponents, such as Ca, Fe, and Al. Breakthrough curves for these components are characterized by sharp edges. The compound of La and Ce (as markers of all other REM) behave similar to nitric acid. Sorption and desorption of La and Ce start with other metals, but go slower, with more flat breakthrough curves. At the regeneration stage, well-defined concentration peaks of Ce and, in particular, La are also observed (together with the acid peak). It makes it possible to obtain separated solutions that are easier to process further for the isolation of target components: Zr-containing solution without REM, at sorption stage, and the acid, cleaned of other metals but enriched with La and Ce, at the regeneration stage. This phenomenon was observed for the first time and requires further study, but now can be regarded as a useful feature that allows carrying out a partial separation of the components in liquid concentrates on the basis of nitric acid.

5.3.5 Pretreatment of Samples for Elemental Analyses

In addition to technological opportunities, consider an example of sample preparation in elemental analysis, in particular, for the efficient reduction of the acidity of solutions produced by the autoclave or microwave decomposition of difficultly soluble materials, such as geological mineral samples or biological tissue like bones. Removal of excess acid from acidic solutions is important in two cases: use of analytical instruments such as ICP MS or ISP ES, for which the samples should not be strongly acidic, and when using combined analytical schemes with preconcentrations where the preparation of solutions is followed by a sorption. In practice, various conventional methods are applied to separate excess acids, for example, controlled evaporation or neutralization. The first of these is characterized by high labor intensity, and the second by an increase in the overall mineralization of the sample, which is also undesirable.

The experimental results in Figure 5.10 demonstrate the prospects for using the NewChem method for de-acidification of solutions before their elemental analysis. A model solution of bone decomposition with nitric acid in autoclave was used in the experiment.

All the concentration curves for microcomponents appear together and are superimposed one another, which is why only the curve for Co is shown. Two features can be noted: (a) the microcomponent is partially sorbed and contained in the first fractions of effluent in lower concentration than the influent and (b) the fractions that correspond to the interval of volume between the vertical lines in the figure can be used for analyses because they are almost free of acid and contain the analyte in concentrations very close to the initial sample.

We do not give the details here, but there is an article[14] that describes the details of this analysis and the results of the reproducibility of the analysis of an actual bone mineralization in repeated experiments on sorption and desorption. The results showed that with tiny columns filled with organic phase, we can not practice consuming techniques and work with small volumes of samples at significantly reduced durations of analytical experiments.

5.4 ION EXCHANGE PROCESSES IN COLUMNS WITH TWO LIQUID PHASES

5.4.1 Simple Ion Exchange Process

It is interesting to understand how the presence of an additional liquid phase in a column and the absence of the inter-particle space in sorption bed affect the results of simple ion exchange process. Figure 5.11 shows the concentration breakthrough curves for the exchange of K^+ and NH_4^+ ions in experiments carried out entirely under the same conditions, except that a column filled with decanol was used in the experiment described in Figure 5.11b. A 1 M solution of ammonium nitrate has been used to displace K^+ from the resin phase of strong acid cation exchanger KU-2 (styrene–divinylbenzene (8%) cationic resin of gel type with sulfonic functional group, manufactured in Russia). Figure 5.11a describes the traditional IEX experiment and shows that at first, water from the porosity space and free volume under the bed leaves the column. This stage continues until the effluent volume indicated by the vertical line is achieved. Then, the salt solution of the displaced ion passes; finally, the influent component increases until equilibrium is reached. In experiment B, the displaced solution with the equilibrium concentration of K^+ (close to NH_4^+ content in the influent) appears immediately. A small difference in these concentrations is related to the small volume of water initially contained in the ion exchange bed. Comparison of the results shows that the use of two immiscible liquid phases in the processes does not affect the total exchange capacity. Under other equal conditions, the new approach reduces the duration of the working cycle, thereby raising the efficiency of the ion exchange column by increasing the number of cycles within the same work period.

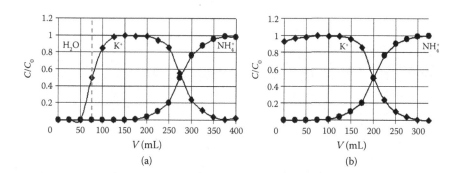

FIGURE 5.11 Concentration histories of K^+ and NH_4^+ in the processes of changing the ionic form of strong acid cationic resin with the use of (a) the standard technique and (b) the NewChem method.

When comparing the results of the experiments presented earlier for AR to those shown by the ion exchange test, the question arises: why does the displaced component in the latter case leave the column immediately with the concentration almost equal to the equilibrium one, while in the above cases, some "likeness" of free volume is observed? It is easy to estimate the said volumes for the different experimental cases described above, and they exceed the volumes of water, which can remain in columns filled with the organic phase (e.g., in the form of thin films coating each resin bead).

This discrepancy can be explained by the fact that in very concentrated solutions, in addition to AR, some retardation of salts also takes place. Such an effect has been proved experimentally and described previously.[31]

5.4.2 Conversion of Potassium Chloride into Sulfate

Here and in the following, we will describe ion exchange processes relevant to the practices, in which highly concentrated solutions are used and which are complicated by accompanied processes, such as ion exchange supersaturation (IXISS effect).[10]

Consider the effect of using the NewChem method by the example of the conversion of potassium chloride to potassium sulfate. In the case of the ordinary IEX technique, two variants of the process can be used.[32] For the first one, a strong base anion exchanger in a stationary bed is transformed from initial chloride form into sulfate using sodium sulfate of any concentration, and then, the solution of potassium chloride close to saturation is passed through a column in the upward direction. In the second variant, a strong acid cationic resin in sodium form is transformed into potassium using the concentrated solution of potassium chloride, and then the nearly saturated solution of sodium sulfate is passed through a column, in the same way, in the upward direction.

1. *Cation exchange process:*

$$R-SO_3Na + KCl \Leftrightarrow R-SO_3K + NaCl \quad\quad\quad (5.8)$$

$$2R-SO_3K + Na_2SO_4 \Rightarrow 2R-SO_3Na + K_2SO_4\downarrow \quad\quad\quad (5.9)$$

2. *Anion exchange synthesis:*

$$2R-N(CH_3)_3Cl + Na_2SO_4 \Leftrightarrow [R-N(CH_3)_3]_2SO_4 + 2NaCl \quad\quad\quad (5.10)$$

$$[R-N(CH_3)_3]_2SO_4 + 2KCl \Rightarrow 2R-N(CH_3)_3Cl + K_2SO_4\downarrow \quad\quad\quad (5.11)$$

In both variants, at the second stage of the processes, during the regeneration of resins into initial ionic forms, a parallel process takes place: a relatively stable supersaturated solution of potassium sulfate is formed inside the bed.[32] From this solution, while leaving the ion exchanging bed, a solid product crystallizes spontaneously.

After insulation of the crystals of potassium sulfate, the next portion of sodium sulfate (in an amount strictly equivalent to the product output) is added into the residual solution of a mixture of sulfates, and the process continues without any liquid waste. Experience shows that the problem of accumulation of sediment over the sorption bed arises often; this effect is the reason as to why the process goes out of the steady-state regime of repeated cycles of sorption–regeneration. When using the proposed method with the direction of flow of regenerating solution from the top down, and at the organization of the process without free volume under the ion exchanging bed, the crystallization in an effluent occurs only outside the column. This eliminates the causes that prevent the formation of a stable steady state. Concentration curves of the components for one of the completely recurring cycles for the variant with the cationic resin are shown in Figure 5.12.

It was interesting to study experimentally the mechanism of potassium sulfate crystallization in the resin bed at the IXISS-based synthesis of K_2SO_4. The test experiment was carried out analogously to the technique proposed earlier[2]: a certain volume of a concentrated Na_2SO_4 solution was passed through the column with KU-2 × 8 cationic resin in the K-form until the formation of a zone of a supersaturated K_2SO_4 solution in the resin bed. Then, the displacing solution flow was stopped. After the completion of K_2SO_4 crystallization in the interstitial space of the column, the resin was removed from the column together with potassium sulfate crystals, separated from crystals, rinsed with a small portion of water, and dried. A dry resin sample was examined by X-ray diffraction analysis, which demonstrated the absence of any crystals inside the amorphous resin phase. The data obtained showed the

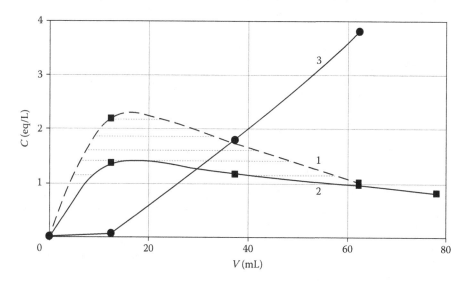

FIGURE 5.12 Concentration histories of K^+ and Na^+ in steady-state process of obtaining potassium sulfate. 1: Concentration of K^+ in supersaturated solution. 2: Concentration of K^+ in residual (saturated) solution after the spontaneous crystallization of K_2SO_4. Bed volume: 40 mL of KU-2 in initial K-form. Influent: saturated solution of Na_2SO_4 K_2SO_4 mixture. Flow rate: 2 bed volumes per hour (BV/h). Organic phase: nonanoic acid.

TABLE 5.7

Quality of Test Product, Potassium Sulfate, Compared with Commercial Reactants of Different Grades (Russian Standards 4145–74)

No.	Mass Fraction (n/L—no less, n/m—no more than)	"Pure"	"Pure for Analyses"	Product
1	Major compound, n/L	97.0	98.0	98.0
2	Insoluble admixtures, n/m	0.02	0.01	< 0.001
3	Nitrates (NO_3^-), n/m	0.004	0.002	–
4	Chlorides (Cl^-), n/m	0.002	0.001	0.0006
5	Ammonia (NH_4^+), n/m	0.004	0.002	0.001
6	Iron (Fe), n/m	0.001	0.0005	<0.002
7	Calcium (Ca), n/m	0.02	0.01	0.005
8	Magnesium (Mg), n/m	0.01	0.004	0.0001
9	Arsenic (As), n/m	0.0004	0.0002	–
10	Sodium (Na), n/m	0.15	0.15	0.004
11	Lead (Pb), n/m	0.001	0.001	–

crystallization of the supersaturated solution only in the solution phase. It can be proposed for the process in columns with two liquid phases that a supersaturated (or colloidal) solution in the bed is more stable because inside the thin film coating the resin beads, it is more difficult for crystal nucleuses to reach critical sizes for spontaneous growing.

Another feature of the IXISS-based conversion of salts is the possibility of directly obtaining a high-grade product without additional processing. Table 5.7 shows data of the chemical analysis of a product prepared by the proposed method (after filtering, washing with saturated solution of K_2SO_4, and drying). The data demonstrate the quality between the grades "Pure" and "Pure for Analyses."

5.4.3 Desorption of Iodine after Its Recovery from Concentrated Chloride Brines

One can ask whether it is possible to create a process in which the presence of an organic phase in the bed of resin improves the process by shifting the equilibrium distribution into a desirable direction, for example, a process in which the organic liquid would play not only the role of an inert phase, but could also be an additional extractant for the target component. Such a process can be illustrated by the extraction of iodine from chloride brines.

Ion exchange extraction of iodine from natural highly mineralized water (pretreated with acid and chlorine) with the use of strong base anionic resin is widely applied. At any sorption stage of recurring working cycles of such a process, the selective sorption of iodine on chloride form of the exchanger usually leads to the formation of its $(I_2)Cl$ form. The desorption stage is often performed with the use of reducing chemical agents like sodium bisulfide mixed with a concentrated solution of sodium chloride to desorb reduced iodine in the form of sodium iodide and regenerate the chloride form.

The alkaline method of iodine desorption is also known, but it has not yet got serious application for practice. The treatment of enriched resin with alkali gives mostly the exhausted exchanger in (I)Cl form, with one atom of iodine, remaining sorbed, per one functional group.[33] It does not provide complete desorption of iodine in each working cycle, and the efficiency of resin is relatively low. The repetitive processes for the sorption and desorption stages for alkaline method can be described as follows:

$$2R-Cl(I) + I_2 = 2R-Cl(I_2) \tag{5.12}$$

$$6R-Cl(I_2) + 6NaOH = 6R-Cl(I) + NaIO_3 + 5NaI + 3H_2O \tag{5.13}$$

The reaction in Equation 5.13 shows the process of disproportionation, which takes place at iodine desorption in alkali media.

When using a very high concentration of the desorbing agent $-$NaOH, the described IEX process can give a supersaturated solution of $NaIO_3$, which spontaneously crystallizes in effluent.[33] It would be interesting to use such a process with the possibility of obtaining not only the highly concentrated iodine-containing product, but also direct insulation of a pure sodium iodate from this product without its further chemical processing. Some of the results of the experimental test of such an opportunity are illustrated in Figure 5.13. To perform the experiments, (I_2)Cl form of strong base anion exchanger AV-17 was preliminarily prepared at bath conditions with the use of calculated amounts of resin and reactants: KI, KIO_3

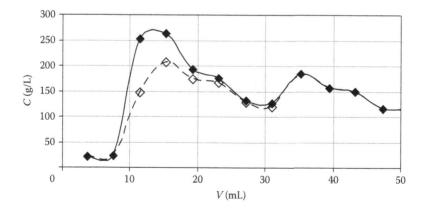

FIGURE 5.13 Concentration curves for the alkaline desorption of iodine with the use of convenient column with sorption bed in aqueous media. Solid line: total concentration of iodine in supersaturated solution. Dashed line: concentration of iodine in residual (saturated) solution after the spontaneous crystallization of $NaIO_3$. Bed volume: 40 mL of AV-17 in initial (I_2)Cl form. Influent: 4.5 M solution of NaOH. Flow rate: 2 bed volumes per hour (BV/h).

and HCl. In the figure, the concentration curves of iodine desorption using 4.5 M NaOH (from the bed, which contained water in free volume and inter-particle space) have a complex shape with two peaks. This is understandable when we take into account the fact that the selectivity properties of an anion exchanger to iodate and iodide formed during disproportionation are extremely different. Sodium iodate leaves the bed much faster than iodide, and area around the first peak on the breakthrough curve accords to the concentration zone enriched with iodate. Total concentration of desorbed components in this zone achieves maximally 270 g/L in terms of iodine content. This also creates a supersaturated solution of worse soluble sodium iodate, which spontaneously crystallizes from the effluent within 1–2 h after leaving the column.

Precision chemical analyses and balance calculations for the experiment described show:

- Forty milliliters of initial $(I_2)Cl$ form of the resin used contained 1.56 g of iodine.
- The total amount of iodine desorbed is equal to 0.75 g (48% of the initial content).

Such a balance with good accuracy corresponds to the process described by Equation 5.13.

Figure 5.14 demonstrates the results obtained in experiments carried out at the conditions practically analogous to those described above, except that the ion exchange bed in the initial $(I_2)Cl$ form has been filled with nonanoic acid.

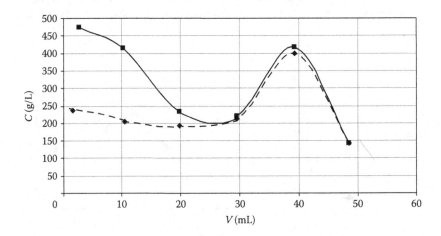

FIGURE 5.14 Concentration curves for the alkaline desorption of iodine with the use of column with two liquid phases. Solid line: total concentration of iodine in supersaturated solution. Dashed line: concentration of iodine in residual (saturated) solution after spontaneous crystallization of $NaIO_3$. Bed volume: 40 mL of AV-17 in initial $(I_2)Cl$ form. Influent: 4.5 M solution of NaOH. Flow rate: 2 bed volumes per hour (BV/h). Organic phase: nonanoic acid.

We can compare the results of two experiments and see that in the last case, a highly concentrated solution leaves the column immediately without free volume zone. The concentration of iodine in the first fractions of effluent achieves 475 g/L; the quantity of sodium iodate, which crystallizes from its supersaturated solution, is much more than in the first (convenient) desorption experiment.

Balance calculations for the experiment with organics show:

- Forty milliliters of the initial $(I_2)Cl$ form of the resin used contained 1.56 g of iodine.
- The total amount of iodine desorbed is equal to 1.48 g (95% of the initial content).

Such a balance corresponds to the process of complete exhaustion of preliminarily enriched resin. The repetitive cyclic processes of sorption–desorption (regeneration) can be described as follows:

$$R-Cl + I_2 = R-Cl(I_2) \qquad (5.14)$$

$$3R-Cl(I_2) + 6NaOH = 3R-Cl + NaIO_3\downarrow + 5NaI + 3H_2O \qquad (5.15)$$

The NewChem process is much more efficient, and we can propose that the organic phase in the process plays the role of additional extractant. Supporting this hypothesis, we can write the equilibriums between aqueous and organic* phases as given below:

$$IO_3^- + 5I^- + 6OH^- \Leftrightarrow 3I_2 \Leftrightarrow 3I_2^* \qquad (5.16)$$

In terms of multilayer dynamics with flow direction from the top to the bottom of the column, the process of desorption, disproportionation, and iodine back formation, followed by its extraction into the organic phase, can take place in any lower layer, while the re-extraction of iodine into the water phase and its disproportionation and displacement (additionally to the main desorption process) can occur in the corresponding upper layers. Thus, the presence of the organic phase enabling the extraction of a displaced component can shift the equilibrium of the desorption stage into a desirable direction and increase the efficiency of ion exchange process.

5.5 CONCLUSIONS

A new approach to the organization of mass transfer processes in ion exchange and sorption columns is proposed. The approach consists in the fact that the columns loaded by granulated resins are additionally filled with organic liquids immiscible with aqueous solutions. In this case, the free volume in the column and inter-particle space in the sorption bed are practically excluded. In the example of a number of topical processes for separating ionic mixtures, the proposed method is tested and the results of laboratory or pilot experiments are given. These processes involve processing of liquid concentrates of aluminum leaching from alumina-containing ores; purification of working electrolytes for copper refining; purification of industrial phosphoric acid with separation of REM compounds; processing of REM- and Zr-containing nitrate concentrate produced at eudialyte decomposition; pretreatment of samples for elemental analyses; ion exchange conversion of potassium chloride into sulfate; and desorption of iodine after its recovery from concentrated chloride brines. For the majority of the processes, organics play the role of the inert phase and do not affect the equilibrium of separation. The advantages of this new approach, which makes it possible for the decrease in the duration of cyclic processes, reducing the longitudinal dispersion effects in dynamics in columns, reduction in special requirements for facilities and their cheapening, and the opportunity to work with supersaturated solution and colloidal systems, are demonstrated for two classes of processes: AR and ion exchange.

ACKNOWLEDGMENTS

The authors are grateful to the Foundation "Leader" of the Russian Venture Company and the Scientific and Industrial Enterprise Joint Stock Company "Radiy" for financially supporting the investigations.

REFERENCES

1. Hatch, M. J., Dillon, J. A. Acid retardation. A simple physical method of separation of strong acids from their salts. *I&EC Process Design Dev.* **1963**, *2*, 253–263.
2. Brown, C. J. *Fluid treatment method and apparatus.* US Patent No. 4673507.
3. Sheedy, M. 1998. Recoflo ion exchange technology. In *Proceedings of the TMS Annual Meeting,* San Antonio, TX, 1987.
4. Ferapontov, N. B., Parbuzina, L. R., Gorshkov, V. I., Strusovskaya, N. L., Gagarin, A. N. Interaction of cross-linked polyelectrolytes with solutions of low-molecular-weight electrolytes. *React. Funct. Polym.* **2000**, *45*, 145–153.
5. Davankov, V. A., Tsyurupa, M. P., Blinnikova, Z. K., Pavlova, L. Self-concentration effects in preparative SEC of mineral electrolytes using nanoporous neutral polymeric sorbents. *J. Sep. Sci.* **2009**, *32*, 64–73.
6. Knudsen, K. C. The production of NPK fertilizers by ion exchange. *J. Appl. Chem. Biotechnol.* **1974**, *24*, 701–708.
7. Rendueles, M., Fernandez, A., Diaz, M. Coupling of ion exchange with industrial processes. Application in fertilizer production and modeling of the key elution step. *Solv. Extr. Ion Exch.* **1997**, *15*, 143–168.
8. Khamizov, R. Kh., Mironova, L. I., Tikhonov, N. A. Recovery of pure magnesium compounds from seawater by the use of IXISS-Effect. *Sep. Sci. Technol.* **1996**, *31*, 1–20.
9. Khamizov, R. Kh., Tikhonov, N. A., Myasoedov B. F. On general character of the IXISS-effect. *Dokl. Phys. Chem.* **1997**, *356*, 216–218.
10. Muraviev, D. N., Khamizov, R. Kh., Tikhonov, N. A. Ion exchange isothermal supersaturation: Concept, problems, and applications. In *Ion Exchange and Solvent Extraction. A Series of Advances*, SenGupta, A., Marcus, Y., Eds. Marcel Dekker, New York, 2004, pp. 119–210.
11. Khamizov, R. Kh., Ivanov, V. A., Madani, A. A. Dual-temperature ion exchange: A review. *React. Funct. Polym.* **2010**, *70*, 521–530.

12. Kuznetsov, I. A., Gorshkov, V. I., Ivanov, V. A., et al. Ion exchange properties of immobilized DNA. *React. Polym.* **1984**, *3*, 37–41.

13. Khamizov, R. Kh., Krachak, A. N., Podgornaya, E. B., Khamizov, S. Kh. The method for carrying out sorption mass-exchange processes, apparatus for its implementation, industrial plant for separation of the components of inorganic water solutions and apparatus for the insulation of organic liquids from water solutions. Patent of Russian Federation No. 2434679, 2011.

14. Khamizov, R. Kh., Krachak, A. N., Podgornaya, E. B., Khamizov, S. Kh. Method and apparatus for conducting mass transfer processes. European Patent EP 2581134 A1, 2013.

15. Khamizov, R. Kh., Krachak, A. N., Podgornaya, E. B., et al. Method of mass transfer processes and the designated apparatus. US Patent Application Publication No.: US 2013/0146543 A1, 2013.

16. Khamizov, R. Kh., Krachak, A. N., Khamizov, S. Kh. Separation of ionic mixtures in columns with two liquid phases. *Sorp. Chromatogr. Process.* **2014**, *14*, 14–23.

17. Podgornaya, E. B., Krachak, A. N., Khamizov, R. Kh. Application of sorption method for the separation of acids and their salts in systems with two liquid phases for the solution of sample preparing problems in elemental analysis. *Sorp. Chromatogr. Process.* **2011**, *11*, 99–110 (in Russian).

18. Khamizov, R. Kh., Krachak, A. N., Khamizov S. Kh. Separation of ionic mixtures in sorption columns with two liquid phases. In *International Conference on Ion Exchange (IEX 2012). Extended Abstracts*, Cox, M., Ed. Queens' College Cambridge, University of Hertfordshire, 2012, pp. 71–72.

19. Tsyurupa, M., Blinnikova, Z., Davankov, V. Ion size exclusion chromatography on hypercrosslinked polystyrene sorbents as a green technology of separating mineral electrolytes. In *Green Chromatographic Technique*, Inamuddin, A. M., Ed. Springer Science+Business Media, Dordrecht, 2014, pp. 19–28.

20. Glotova, E. A., Tikhonov, N. A., Khamizov, R. Kh., Krachak, A. N. Mathematical modeling of a sorption process for the retention of acid from a solution. *Moscow Univ. Phys. Bull.* **2013**, *68*, 65–70.

21. Sidelnikov, G. B., Tikhonov, N. A., Khamizov, R. Kh., Krachak, A. N. Modeling and research of acids and salts sorption separation process in the solution. *Matem. Mod.* **2013**, *25*, 3–16.

22. Khamizov, R. Kh., Vlasovskikh, N. S., Moroshkina, L. P., Khamizov, S. Kh. Scientific grounds and prospects for closed-circuit processing of alumina-containing raw materials with the decomposition of salt and acid-salt types. In *ICSOBA—2014, Proceeding Book of 32nd International Conference @New Challenges of Bauxite, Alumina and Aluminium Industry and Focus on China*, Zhengzhou, ZRI CHALCO, 2014, pp. 249–258.

23. Dorre, E., Hubner H. *Alumina: Processing, Properties, and Applications*, Springer, London, 2011.

24. Comyns, A. E. *Encyclopedic Dictionary of Named Processes in Chemical Technology.* CRC Press LLC, Boca Raton, FL, 2000.

25. Schrödter, K., Bettermann, G., Staffel, T., et al. Phosphoric acid and phosphates. In *Ullmann's Encyclopedia of Industrial Chemistry,* ed. advis. board: G. Belussi, M. Bohnet, J. Bus et al., Wiley-VCH, Weinheim, 2008, V.26, pp. 679–724.

26. Khamizov, R. Kh., Krachak, A. N., Gruzdeva, A. N., et al. Concentration and insulation of REE from industrial extractive phosphorous acid by sorption method. *Sorp. Chromatogr. Process.* **2012**, *12*, 29–39 (in Russian).

27. Vlasovskih, N. S., Khamizov, S. Kh., Khamizov, R. Kh., et al. Extraction of impurities: REM and other metals, from phosphoric acid. *Sorp. Chromatogr. Process.* **2013**, *13*, 605–617 (in Russian).

28. Khamizov, R. Kh., Krachak, A. N., Gruzdeva, A. N., et al. Separation of REE from industrial extractive phosphorous acid by sorption method. RF Patent Application, Publication No. 2013133099/02(049460), 2014.

29. Evtushenko, A. V., Pavlukhina, L. D., Yakusheva, A. M., et al. Influence of concentration of acid and its salt on sorption-desorption processes in the method of Acid Retardation. *Sorp. Chromatogr. Process.* **2013**, *13*, 313–321 (in Russian).

30. Zajtsev, V. A., Gruzdeva, A. N., Gromyak, I. N., Sedych, E. M., Roshcina I. A., Kogarko L. N. Eudialyte decomposition rate in nitric acid. *Exp. Geochem.* **2014**, *2*, 296–99 (in Russian).

31. Krachak, A. N., Khamizov, R. Kh. Basic regularities of electrolyte separation in the method of Acid Retardation. *Sorp. Chromatogr. Process.* **2011**, *11*, 77–88 (in Russian).

32. Muraviev, D. N., Khamizov, R. Kh., Fokina, O. V., Tikhonov, N. A., Krachak, A. N., Zhiguleva, T. I. Clean ion-exchange technologies. II. Recovery of high purity magnesium compounds from seawater by ion-exchange isothermal supersaturation technique. *Ind. Eng. Chem. Res.* **1998**, *37*, 2496–2501.

33. Krachak, A. N., Khamizov, R. Kh., Pinaeva, I. P., Bychkov, A. M. Study of ion exchange process with the formation of supersaturated solution of iodine. *Sorp. Chromatogr. Process.* **2008**, *8*, 11–22 (in Russian).

6 Sensing of Toxic Metals Using Innovative Sorption-Based Technique

Prasun K. Chatterjee and Arup K. SenGupta

CONTENTS

A variety of heavy metals is on the list of priority pollutants because of their toxic effect on human health and living organisms. Sensing the presence of toxic heavy metals (e.g., lead, copper, zinc, and nickel) in aquatic environments including drinking water is exceedingly important and often the first step to determine if any corrective action including treatment is necessary. Although sorption chemistry has been widely applied for the removal of various toxic metals from water, a sorption-based technique has seldom been studied as a stand-alone approach for detecting toxic metals in water. The present chapter reports a new sorption-based technique for sensing toxic metals such as lead, copper, zinc, and nickel in water using pH as the sole surrogate indicator through the use of novel sorbent materials. An environmentally benign, inexpensive hybrid inorganic sorbent material forms the heart of the process. This hybrid material is essentially a granular composite of calcium–magnesium–silicate ($Ca_2MgSi_2O_7$) synthesized through a rapid thermal fusion technique. Two varieties of sorbent materials, that is, hybrid inorganic material (HIM) and hybrid inorganic material next generation (HIM-X), are used. A column-run using sorbent materials (i.e., HIM/HIM-X) for water containing common electrolytes such as sodium, calcium, chloride, and sulfate exhibits a consistent alkaline pH (~9.0) at the column exit because of the steady release of hydroxyl ions (OH^-) from slow hydrolysis of calcium–magnesium–silicate. In contrast, the presence of trace concentrations of toxic metals (e.g., lead <100 μg/L) in feedwater shows a significant drop in pH (>2 units) in the exiting solution under otherwise identical conditions. A sharp change in pH of the exit solution signals the presence of toxic metal(s) in the feed, which can also be detected by color changes of the indicator solution. At this point, upon withdrawal of toxic metal from the feed, the exit pH again rises to the alkaline domain, which confirms that the exit pH is responsive to fluctuations of toxic metal concentration in the feed. The slope of the pH curve (i.e., $-dpH/dBV$, where BV is the bed volume of solution passed) shows a sharp and distinctive peak for different toxic metals. Since pH is the sole detecting parameter, experimental techniques validate overcoming the interference of common buffering species such as bicarbonate/carbonate, and phosphate and natural organic matter (NOM) containing weak-acid anionic ligands. While the interference of bicarbonate is avoided through adjustment of the inlet pH, the interference of phosphate and NOM is overcome using a hybrid anion exchanger (HAIX) prior to HIM/HIM-X bed. In HAIX, hydrated ferric oxide (HFO) nanoparticles, dispersed within the polymeric phase, selectively remove phosphate or other weak-acid anionic ligands (e.g., phosphate, fulvate in NOM). This technique validates the detection of lead and zinc in synthetic water, tap water, and river water. The sensitivity of this technique for sensing an ultra-low concentration (<10 μg/L) of toxic metal in water is verified with the aid of a preconcentration analytical approach. Except for the pH meter or indicator solution, no other sophisticated instruments or chemicals are required for detection purposes. This technique seems to have immense potential for field applications for identifying toxic metal contamination in water in both the developed and the developing world.

6.1 INTRODUCTION

6.1.1 NEED FOR SENSING OF TOXIC HEAVY METALS

Contamination of drinking and natural water sources by various toxic heavy metals such as lead, copper, nickel, cadmium, and zinc has become a major issue over the last few decades. Since the era of industrial development, metal cycles in the environment have been profoundly modified on a regional and global basis. Besides natural and/or geo-chemical processes, metallurgical, electroplating, and other anthropogenic and industrial applications cause ingress of heavy metals into the environment. In addition, acid rains and the use of surfactants and chelating compounds have increased the mobility of heavy metals through the environment.[1] Some of these metals are micronutrients (e.g., copper, zinc) that are essentially needed at trace concentrations for unhindered metabolism, but all of them pose serious health risks, if ingested, beyond a certain limit.[1,2] As a result, all of them are environmentally regulated. Table 6.1 exhibits the maximum contaminant levels (MCLs) of different heavy metals and metalloids (e.g., arsenic, antimony) of concern in drinking water according to the U.S. Environment Protection Agency (EPA).[3] An operationally

TABLE 6.1
MCLs for Regulated Heavy Metals in Drinking Water

Heavy Metal	MCL (mg/L)	Category	Health Effects	Source
Zinc	5.0[a]	Essential nutrient	Excess zinc can cause epi-gastric pain and lethargy and suppresses copper and iron absorption	Erosion of natural deposit, pigment run-off, steel mill waste
Copper	1.3	Essential nutrient	Gastro-intestinal disorder for short-term exposure, liver and kidney damage for long-term exposure	Corrosion of household plumbing, erosion of natural deposits
Chromium (total)	0.1	Essential nutrient	Allergic dermatitis, hexavalent form is considered as carcinogen	Discharge from steel and pulp mills, erosion of natural deposits
Nickel	0.07[b]	Essential nutrient	Allergic dermatitis, probable carcinogenic	Erosion of natural deposits, mines run-off
Cobalt	No specific guide lines[c]	Essential nutrient	Contact dermatitis, cobalt in some compounds are probable carcinogenic. More toxic to aquatic organisms[c]	Erosion of natural deposits, plating wastes, run-off from paints, waste batteries
Lead	0.015	Non-essential	Delays in physical and mental development for children, kidney damage, and high blood pressure for adults	Corrosion of household plumbing, erosion of natural deposits
Cadmium	0.005	Non-essential	Kidney damage	Erosion of natural deposits, run-off from waste batteries, paint, discharge from metal refineries
Mercury	0.002	Non-essential	Kidney damage, nervous system disorder	Erosion of natural deposits, discharge from refineries, run-off from landfills

Source: U.S. EPA, *National Primary Drinking Water Regulations—List of Drinking Water Contaminants and Their MCLs*, U.S. EPA, 2003 (Adapted from http://www.epa.gov/safewater/mcl.html).

[a] Included in national secondary drinking water regulation (EPA).

[b] World Health Organization (WHO) standard (WHO/SDE/WSH/05.08/55).

[c] Co is essential as a component of vitamin B_{12}. Total body burden is estimated as 1.1–1.5 mg. The largest source of exposure is food supply. No studies describe the distribution of Co in humans following inges-tion. For rats, oral LD_{50}s for soluble cobalt compounds have been reported to range from 42.4 to 317 mg/kg of body weight depending on the compound and species tested according to WHO Concise International Chemical Assessment Document 69, 2006, ISBN 9241530693 (http://whqlibdoc.who.int/publications/2006/9241530693_eng.pdf).

simple, inexpensive, and useful technique for sensing trace concentrations of environmentally regulated toxic metals in water is a necessity in many diverse applications for water quality control and mitigation of health hazards.

6.1.2 IMPACT OF HEAVY METALS IN THE ENVIRONMENT

The term "heavy metals" commonly used in specifying groups of metals of environmental and biological significance is rather imprecise and arbitrary. It is often used as a group name of metals and metalloids that have been associated with contamination and potential toxicity. These metals are commonly identified as the transition and post-transition groups of elements in the periodic table, and classified according to Lewis acid behavior in predicting their chemical interactions. Lewis acids are defined as elemental species with a reactive vacant orbital or more precisely as an electron pair acceptor.[4] An assessment of behavior of metal ions as electron acceptors determines the possibility of reaction or complex formation. The classification of metals by their Lewis acidity is shown in Table 6.2, which results in three categories of metals namely, hard acids (class A), soft acids (class B), and borderline acids (intermediate)[5,6] as per hard soft acid base (HSAB) rules. Note that most of the heavy metals of interest fall under the "borderline" and "soft" categories. Borderline and soft cations usually bind strongly with Lewis bases (electron pair donor) forming inner sphere complexes with higher preferences for nitrogen, sulfur, and oxygen electron-donating ligands.

TABLE 6.2
Classification of Selected Metal Cations According to HSAB Rules

Category	Name of Metal Cations	Salient Properties
Class A (hard metals)	Na^+, K^+, Li^+, Ca^{2+}, Mg^{2+}, Al^{3+}, etc.	Spherically symmetric, poor Lewis acid, low polarizability, form outer-sphere complexes with hard ligand containing O donor atoms, most of them are non-toxic at low concentration
Class B (soft metals)	Cu^+, Cd^{2+}, Hg^+, Hg^{2+}, Ag^+, Pd^{2+}, etc.	Spherical not symmetric, strong Lewis acid, high polarizability, high affinity toward S atom containing ligand, very toxic from a physiological view point
Border line metals	Cu^{2+}, Zn^{2+}, Ni^{2+}, Pb^{2+}, Co^{2+}, Fe^{2+}, Mn^{2+}, Fe^{3+}	Spherical not symmetric, strong Lewis acid, relatively high polarizability, form inner sphere complexes with O, N atom containing ligands, except iron and manganese all are toxic

Source: Pearson, R. G., *J. Chem. Educ.*, 45(9), 581–587, 1968; Pearson, R. G., *J. Chem. Educ.* 45(10), 643–648, 1969.

In general, the toxicity of metals increases from hard to borderline and then to soft. Among them, essential elements (e.g., Cu, Ni, Cr) are toxic beyond the beneficial dose; these are beneficial at trace levels but become inhibitory at higher levels. Others with no biological benefit, for example, lead and cadmium, are toxic even at very trace levels, as explained in Figure 6.1. The level of toxicity differs for each metal primarily governed by the chemistry associated with its physiological effects. Other metals in the wrong form also can be toxic. For example, chromium as the Cr^{+3} ion is an essential trace element important for maintaining correct blood-sugar levels, as opposed to Cr^{+6}, which is known to be a human carcinogen.[1,3] The degree of toxicity for a particular metal may vary widely for different species. For example, copper is a micronutrient to human beings, which is essentially needed at trace concentrations to ensure unhindered metabolism but is toxic at higher concentrations. Relative to humans, copper is exceedingly toxic to fish and other aquatic biota. Toxicological tolerance limits of copper in fish and crustaceans are generally 10- to 100-fold lower than those of mammals.[7] Some of these toxic heavy metals are also known for bio-magnification through higher levels of the food chain (e.g., Pb, Hg).[8] The toxic effects of most metals can be traced to their ability to disrupt the function of essential biological molecules, such as proteins, enzymes, and DNA. In some cases, this involves displacing chemically related (similar) metal ions that are required for important biological functions such as cell growth, division, and repair. Lead (Pb^{2+}) can replace calcium (Ca^{2+}) in bone and other sites where calcium is essentially required.[9] Some toxic metals imitate the actions of essential elements in the body, interfering with the metabolic processes and causing illness. For instance, cadmium belongs to the same group as zinc in the periodic table, bears the same

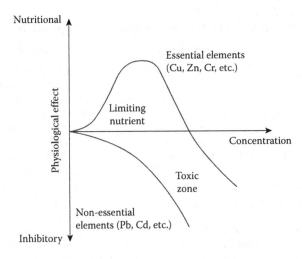

FIGURE 6.1 Nutritional and toxic effects of different heavy metals as a function of concentration on living cells (microorganisms). (From SenGupta, A. K., Ed., *Environmental Separation of Heavy Metals—Engineering Processes*, Lewis, New York, 2002; Stum, W., and Morgan, J. J.: *Aquatic Chemistry: Chemical Equilibria and Rates in Natural Waters*, Third Edition. 1995. Copyright Wiley-VCH Verlag GmbH & Co. KGaA. Reproduced with permission.)

oxidation state (+2), has almost the same size when ionized, and interferes with zinc binding proteins and causes toxicity. In addition, cadmium can replace calcium and magnesium in some biological systems.[9,10]

The relative affinities of these metal ions to form complexes with oxygen-, nitrogen-, and sulfur-containing ligands vary widely. The higher preference of soft and borderline Lewis acids for nitrogen and sulfur species causes binding of these metal cations with sulfhydryl groups of proteins of the cells. Sulfhydryl groups (–SH), which are the active sites of protein, when blocked with heavy metals, the result is a disruption of essential physiological processes.[11] Prompted by these phenomena, Nieober and Richardson recommended classifying toxic metals by their relative complex-forming abilities with O-, N-, S- containing ligands because such affinities are the primary determinant of toxicity caused by these metals.[12] Understanding bioavailability is the key to assessing the potential toxicity of metallic elements and their compounds. Bioavailability further depends on biological parameters and on the physicochemical properties of the metals, their ions, and compounds. Contamination of toxic metals with water is a function of solubility. Insoluble metals and their compounds often exhibit negligible toxicity. Unlike organic pollutants, metals do not decay and thus pose a different kind of challenge for remediation. Commonly, different physicochemical processes such as precipitation, reduction–oxidation reactions, ion exchanges, membrane separations, and various sorption processes are applied for the separation and containment of toxic metals and their compounds.[1,2,11]

6.1.3 SENSING OF HEAVY METALS—DIFFERENT APPROACHES: SHORTCOMINGS AND CHALLENGES

Undoubtedly selective sensing of environmentally regulated heavy metals in water is important. The conventional analytical instruments, which include ion-selective electrodes, atomic absorption spectrometer (AAS), and inductively coupled plasma mass and atomic emission spectrometer, are sophisticated but expensive instruments, and are useful only under proper laboratory settings. For rapid and selective identification of toxic heavy metals in water, several scientific methods are either being investigated or are in the process of development. A couple of such techniques are as follows: (i) Electrochemical sensing through measurement of potential difference between electrodes.[13,14] (ii) Electrochemical analysis using surface polarization.[15] (iii) Bio-sensing using the principle of protein-based metal-specific sensing.[16,17] (iv) Optical sensing through the use of optically sensitive organic compounds. The optical property is measured through absorbance or reflectance spectroscopy.[18] (v) Fluorescent technique using metal-specific fluorescent compounds (e.g., chromophores). The use of conjugated chromophores of varying lengths has resulted in the identification of different metal ions.[19,20] (vi) Nanotechnology-based sensor exploiting quantum phenomena (e.g., conductance quantization, quantum tunneling) at the nanoscales.[21,22] However, all of these techniques have their limitations in terms of sample preparation, interferences with different ionic species, reproducibility of results, and multiple uses. In spite of abundant research data, commercialization and large-scale use in the public domain of these techniques are limited as of now.

6.1.4 SORPTION AS A TOOL FOR SENSING

Sorption technology is regarded as one of the most useful tools for removing different toxic heavy metals from water. In a typical sorption-based water purification process, contaminants (e.g., heavy metals) from the water phase are separated and concentrated on a solid phase called a sorbent. Sorption reactions are primarily intermolecular interactions between the solute and the solid phase. In a qualitative sense, sorption of molecules and ions onto a solid surface is the result of reduction of interfacial tension. For example, the interfacial tension of an oxide surface (e.g., quartz) has a relative maximum value at its point of zero charge; its interfacial energy decreases at pH values above and below the pH at point of zero charge or pHzpc.[11,23] Several factors such as the charge of a solute, the hydrated ionic radius, the size and type of a molecule and ions, solution pH, ionic strength, surface area of a sorbent, type/nature of sorbent, and temperature all influence the sorption processes.

Inorganic sorbent materials can act as ion exchangers if their structure bears an excess electrical charge. Acquiring of excess charge is possible by the following ways[24]:

1. In the lattice M^{z+}, ions are replaced by $M^{(z-1)+}$, resulting in excess negative charge on the sorbent balanced by mobile cations.
2. The surface of the sorbent material contains functional groups that get ionized due to protonation or de-protonation. The resulting electrical charges are balanced by counterions.

Zeolites, the aluminosilicate $[Na_x (AlO_2)_x (SiO_2)_y \cdot zH_2O]$ members of microporous inorganic sorbents, are often used for removing cations in water purification through ion exchange and belong to the first category. Some tetravalent Si atoms in the lattice are replaced by trivalent Al atoms producing excess negative charge, which are compensated by mobile cations as explained in the two-dimensional and three dimensional representations in Figure 6.2. Such ion exchange is the result of weak

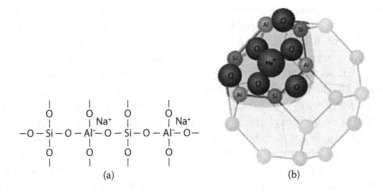

(a)   (b)

FIGURE 6.2 Schematic representation of cation sorption by aluminosilicates (zeolite). (a) Two-dimensional scheme and (b) three-dimensional structure and location of sorbed cation (Me^+ represents sorbed metal cation). (From Wolfgang, H. H., *Ion Exchange: Exchangers, Fundamentals, Applications, Technology,* Forschungszentrum Karlsruhe Institute of Technical Chemistry, 2008.)

electrostatic interactions and does not offer any selective removal of trace heavy metals from the background of large concentrations of common cations (e.g., Na^+, Ca^{2+}). Due to the well-defined pore structure of zeolites, they are used as molecular sieves as well as for separation based on a size exclusion process.[24]

Hydrated oxides of many polyvalent metals, namely, Fe(III), Al(III), Zr(IV), Si(IV), and Ti(IV), are well-recognized sorbents for successful removal of toxic metals or anionic ligands from water depending upon the solution pH.[25–28] In the presence of water, the surfaces of these oxides are covered with hydroxyl groups as explained in Figure 6.3. Depending on the solution pH, they may be protonated or de-protonated, providing room for accommodating cations or anions.[11,25,29] For all these materials, the pH value at which the net surface charge is zero is called the point of zero charge or pHzpc. Figure 6.4 shows a plot of surface charge versus pH for some representative inorganic different oxide materials of interest[11] and in Table 6.3 approximate pHzpc of relevant inorganic oxides are mentioned. Figure 6.5

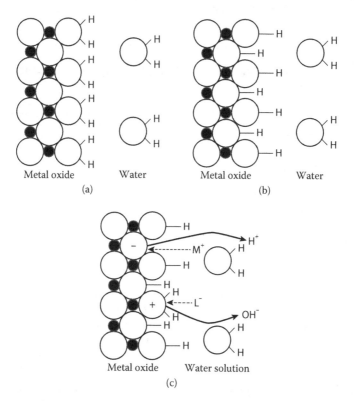

FIGURE 6.3 Schematic representations of surfaces of metal oxides with metal ions (•), oxide ions (O) in the presence of water. (a) Surface metal ions with coordination of H_2O molecules. (b) Dissociative chemisorptions of water molecules. (c) Surface hydroxyl sites on a metal oxides and ion exchange reaction at oxide/solution interface ("M" metal and "L" ligand). (From Stum, W., and Morgan, J. J.: *Aquatic Chemistry: Chemical Equilibria and Rates in Natural Waters,* Third Edition. 1995. Copyright Wiley-VCH Verlag GmbH & Co. KGaA. Reproduced with permission; Tamura, H., et al., *Environ. Sci. Technol., 30,* 1198–1204, 1996.)

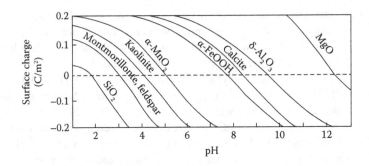

FIGURE 6.4 Effect of pH on approximate surface charge (C m^{-2}) of a few representative inorganic oxide surfaces in contact with water showing approximate pH (pHzpc) with a net surface charge of zero. (From Stum, W., and Morgan, J. J.: *Aquatic Chemistry: Chemical Equilibria and Rates in Natural Waters,* Third Edition. 1995. Copyright Wiley-VCH Verlag GmbH & Co. KGaA. Reproduced with permission.)

TABLE 6.3
Approximate pH at Zero Net Surface Charge (pHzpc) for Different Metal Oxides

Oxide Surface	pHzpc
MgO (periclase)	12.3
α-Al$_2$O$_3$ (corundum)	9.1
α-TiO$_2$ (rutile)	5.8
α-Fe$_2$O$_3$ (hematite)	8.5
HFO (hydrated ferric oxide)	7.7–8.3
Calcite (mainly CaCO$_3$)	8.6
Sepiolite (Mg$_4$Si$_6$O$_{15}$(OH)$_2$ · 6H$_2$O)	7.3
Feldspar (KAlSi$_3$O$_8$ – NaAlSi$_3$O$_8$ – CaAl$_2$Si$_2$O$_8$)	3.4
α-SiO$_2$ (quartz)	2.9

Source: Stum, W., and Morgan, J. J.: *Aquatic Chemistry: Chemical Equilibria and Rates in Natural Waters,* Third Edition. 1995. Copyright Wiley-VCH Verlag GmbH & Co. KGaA. Reproduced with permission.; Dzombak, D. A., and Morel, F. M. M.: *Surface Complexation Modeling: Hydrous Ferric Oxides.* 1990. Copyright Wiley-VCH Verlag GmbH & Co. KGaA. Reproduced with permission.

---

explains that sorption of copper, cadmium, zinc, and nickel cations on hydrated Fe(III) oxide is very good at slightly alkaline or near neutral pH, whereas sorption decreases significantly as the pH goes below 6.0.[29] In comparison, anionic ligand uptake by Fe(III) or Al(III) hydrous oxides is highly favored by low pH conditions and sorption diminishes with an increase in pH as shown in Figure 6.6.[11,29] The solution pH is the master variable that primarily controls the sorption process.

Different previous investigations on heavy-metal separation using engineered sorbent materials reveals that pH can serve as a tool for signaling the presence of

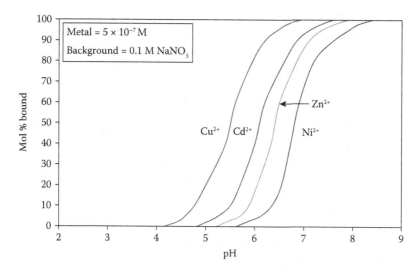

FIGURE 6.5   Extent of different heavy-metal binding on hydrous ferric hydroxide surface as a function of pH with metal concentration in solution of 5×10^{-7} M in 0.1 M $NaNO_3$. (From Dzombak, D. A., and Morel, F. M. M.: *Surface Complexation Modeling: Hydrous Ferric Oxides.* 1990. Copyright Wiley-VCH Verlag GmbH & Co. KGaA. Reproduced with permission; Benjamin, M. M. and Leckie, J. J., *J. Colloid. Interf. Sci., 79,* 209–221, 1981.)

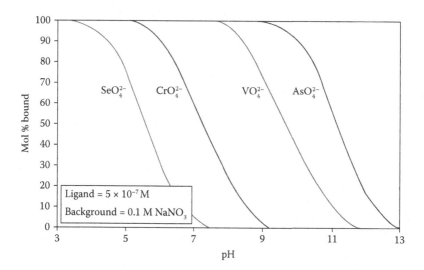

FIGURE 6.6   Binding of ligands (anions) onto hydrous ferric hydroxide surface as a function of pH. Ligand concentration of 5×10^{-7} M in 0.1 M $NaNO_3$ solution. (From Stum, W., and Morgan, J. J.: *Aquatic Chemistry: Chemical Equilibria and Rates in Natural Waters,* Third Edition. 1995. Copyright Wiley-VCH Verlag GmbH & Co. KGaA. Reproduced with permission; Dzombak, D. A., and Morel, F. M. M.: *Surface Complexation Modeling: Hydrous Ferric Oxides.* 1990. Copyright Wiley-VCH Verlag GmbH & Co. KGaA. Reproduced with permission.)

toxic heavy metals. Studies of Gao et al.[30] for separation of heavy metals using synthesized inorganic sorbent prepared from iron oxides and silicates in a fixed-bed column run elucidated a sudden drop in pH and breakthrough of heavy metal in the effluent occurred concurrently. A similar observation relating the drop of pH and simultaneous breakthrough of toxic heavy metals at the exit of the column has been reported by Cortina et al.,[31] Rotting et al.[32] for their studies with caustic magnesia as sorbent, and by Jachova et al.[33] using a calcium-loaded lignite-based sorbent to separate the heavy metals. This phenomenon essentially interprets effluent pH and acts as a surrogate parameter indicating the breakthrough of toxic metals. As an alternative to separating heavy metals through sorption, this chapter focuses on the scientific premise complemented by research investigations for sensing the presence of toxic heavy metals in water through the use of inexpensive, commonly available sorbents or a combination of sorbents.

6.2 CONCEPTUALIZED APPROACH FOR SORPTION-BASED SENSING THROUGH pH CHANGE

6.2.1 COMMONALITIES OF TOXIC HEAVY METALS

The different heavy metals such as zinc, copper, nickel, and lead that are the subjects of interest in the present study, because of the configurations of their electron shells, they all are strong Lewis acids, that is, good electron pair acceptors. Lewis acids are defined as elemental species with a reactive vacant orbital (e.g., vacant 3d orbital for Cu, Zn) or an available lowest unoccupied molecular orbital. In other words, any elemental species with a net positive charge behaves as a Lewis acid because it acts as an electron pair acceptor. The electron clouds of these atoms are more readily deformed by the electric field of other species (e.g., ligands). They have many valence electrons, higher polarizability, and may be visualized as "soft sphere" ions. In general, metals with higher polarizability are found to have increased strength covalent bonding.[4,11,34] Thus, all these metals exhibit strong coordination interactions (i.e., coordinate covalent bonding) as central atoms with a variety of ligands or electron-donating species (Lewis base) containing O, N, and S. All these metal cations have coordination number (C.N.) 4 or more (e.g., Zn^{2+} C.N. 4 and Cu^{2+} C.N. 6). Such metal–ligand interactions are commonly designated as a Lewis acid–base interaction or formation of an inner sphere complex through electron pair sharing. Hard cations such as Na^+, K^+, and Ca^{2+} are spherically symmetric, not easily deformed, poor electron pair acceptors, and interact weakly with these ligands, forming only outer sphere complexes (electrostatic interactions).[4,11,35]

6.2.2 SCIENTIFIC PREMISE FOR SENSING OF TOXIC METALS THROUGH pH CHANGE

The fundamentals of sensing toxic heavy metals through a pH change using an inorganic sorbent material rely upon the following major scientific tenets:

- The ion sorption ability of different surfaces containing hydroxyl or carboxylate groups arises from protonation or de-protonation due to dissociative chemisorption of water molecules at different pHs.[11]

The surface hydroxyl groups, upon de-protonation at pH above pHzpc, offer electron pair donating oxygen atoms (Lewis base) which selectively bind heavy-metal cations such as lead, copper, zinc, and nickel (Lewis acid) through the formation of inner sphere complexes or Lewis acid–base interactions.[11,29,35] Figure 6.7 illustrates the binding of Cu^{2+} onto a de-protonated hydrous oxide surface through forming inner sphere complexes. The oxygen donor atoms of surface functional groups increase the electron density of the coordinated metal ions (Cu^{2+}) and form much stronger complexes than if they were bound by outer sphere mechanisms.

- In contrast, commonly encountered non-toxic alkaline and alkaline-earth metal cations, namely, Na^+, K^+, Ca^{2+}, and Mg^{2+}, form only outer sphere complexes (weak electrostatic or coulombic interactions) as explained in Figure 6.7 with the de-protonated oxygen donor atom of the surface hydroxyl group and are poorly bound to the surface functional groups of hydrous metal oxide surfaces.[11,29,35,36]

- All these toxic metals (M^{2+}) form strong and labile metal–hydroxy complexes at alkaline pH according to the reaction

$$M^{2+}(aq) + nOH^-(aq) \leftrightarrow [M(OH)_n]^{2-n} \tag{6.1}$$

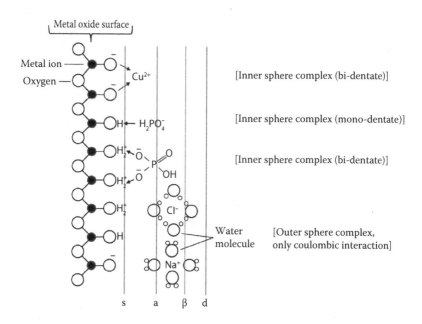

FIGURE 6.7 Schematic illustration of surface functional groups of an hydrous metal oxide surface and their predominant interactions with different solutes showing planes associated with surface hydroxyl groups ("s"), inner sphere complexes ("a"), outer sphere complexes ("β"), and diffuse ion swarm of the double layer ("d"). Lewis acid–base interaction or inner-sphere complex formation is shown by arrows (←). (Adapted from Sposito, G., *The Surface Chemistry of Soils*, 1984, by permission of Oxford University Press.)

On the contrary, the other commonly present innocuous cations, namely, Na^+, K^+, Ca^{2+}, and Mg^{2+}, exhibit very poor affinity for OH^- ions and form only weak metal–hydroxy complexes. It is apparent that the trivalent metals (Fe^{3+}, Al^{3+}) are typically more reactive than divalent ones, but such reactivity is largely offset by the greater insolubility of their corresponding oxides and hydroxides. As a result, other divalent metal cations including Cu^{2+} and Pb^{2+} are more apt to form metal hydroxy complexes in natural water.[37] Thus, the presence of divalent heavy-metal cations in slightly alkaline pH (8–9) will possibly lead to sequestration of OH^- ions due to the formation of metal–hydroxy complexes causing a resultant drop in the solution pH.

6.2.3 CONCEPTUALIZED APPROACH

Conceptually, an adsorbent material that contains surface hydroxyl and/or carboxylate functional groups and remains slightly alkaline (with pH around 9.0) in contact with water can serve as a tool to sense the presence of toxic heavy metals through pH changes[38–41] as illustrated in Figure 6.8. For an idealized sorbent containing surface hydroxyl or carboxylate functional groups, if the surrounding pH remains slightly acidic (pH ~4) in contact with water, the surface functionalities will remain protonated and offer no affinity to any metal cations whatsoever (Figure 6.8a). If a typical water sample containing common cations (Na^+, Ca^{2+}) and anions $\left(Cl^-,\ SO_4^{2-}\right)$ is passed through the bed of conceptualized sorbent material with the surrounding pH 9.0, Na^+ and Ca^{2+} will show very poor affinity for the de-protonated surface binding site and the exit pH remains alkaline as shown in Figure 6.8b. If any dissolved toxic metal (e.g., zinc) is present in the feed solution at trace concentration in addition to the other electrolytes, Zn^{2+} will selectively bind to its surface binding sites through the formation of inner sphere complexes while other electrolytes will rapidly pass through and the exit solution will still remain alkaline as shown in Figure 6.8c. However, with the passage of samples containing Zn^{2+} through the sorbent bed, the surface sites will be gradually exhausted and Zn^{2+} will exit from the bottom of the sorbent column instantaneously, forming complexes with OH^- ions that result in a sharp drop in pH as explained in Figure 6.8d. Thus, the presence of trace zinc in the sample will result in a drop of the exit pH, signaling the presence of toxic metal in the influent. If the heavy metal is withdrawn from the influent, the exit pH will swing back to the alkaline domain, revealing the drop in pH is solely responsible for the toxic heavy metal in the sample influent.

Ideally, a step feed of zinc through this conceptualized sorbent column that maintains a surrounding pH 9.0 in contact with water will show a drop in solution pH at the exit of the sorbent bed with a certain lag time; and upon withdrawal of zinc from the feed the exit pH will turn back to the original (9.0) which has been explained in a conceptualized diagram shown in Figure 6.9a and b. A drop in pH in the exit solution from the sorbent bed will essentially signal the presence of toxic heavy metal in the feed sample, which otherwise will remain alkaline for common electrolytes in water.[38,39]

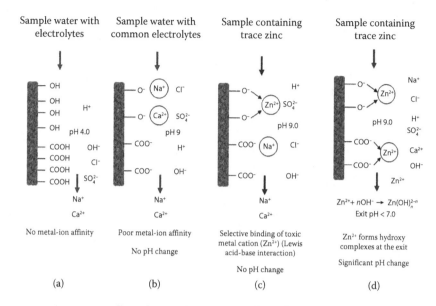

FIGURE 6.8 Illustration of the concept of sensing of toxic heavy metals in water through pH change using the sorption-based approach. (a) At pH ~4.0 with protonated surface functionalities, no affinity for metal cation. (b) Poor sorption affinity for alkaline or alkaline–earth metal cations (Na^+, Ca^{2+}) to de-protonated surface sites at slightly alkaline pH (~9) as a result of weak electrostatic interaction. (c) Selective binding of toxic metal cation (e.g., Zn^{2+}) on de-protonated surface sites through Lewis acid–base interaction at alkaline pH. (d) Metal breakthrough upon saturation of binding sites followed by formation of metal–hydroxy complexes at alkaline pH result in change of aqueous phase pH. (From Chatterjee, P. K., and SenGupta, A. K., *AIChE J*, 55 (11), 2997–3004, 2009; Chatterjee, P. K., *Sensing and Detection of Toxic Metals in Water with Innovative Sorption Based Techniques*, PhD Thesis, Civil and Environmental Engineering, Lehigh University, Bethlehem, PA, 2010.)

6.2.4 GENESIS OF CONCEPTUALIZED SORBENT MATERIAL

The primary challenge in transforming the concept into reality lies in synthesizing the sorbent material that possesses both selective binding sites with oxygen donor atoms for toxic metals and also maintains a near-constant alkaline pH in contact with aqueous solution. Oxides/hydroxides of polyvalent cations such as Fe(III), Al(III), Si(IV), Ti(IV), and Zr(IV) are widely known for selectively binding heavy metals onto their de-protonated surface sites at slightly alkaline pH.[25,27,35] Different types of silicates, for example, kaolinite, sepiolite, illite, vermiculite, and so on, which are commonly referred to as clay minerals, are well documented for sorption of heavy-metal cations in numerous instances.[25,42–44] From the reference of previous studies at Lehigh University for the separation of heavy metals using inorganic sorbent materials[30,45] akermanite (a mineral mainly composed of alkaline metal silicates) has been known to undergo an incongruent dissolution reaction resulting in the release of hydroxyl ions through slow hydrolysis. Subsequent to this background, an HIM with attributes of the "conceptualized sorbent material," that is, binding sites for toxic

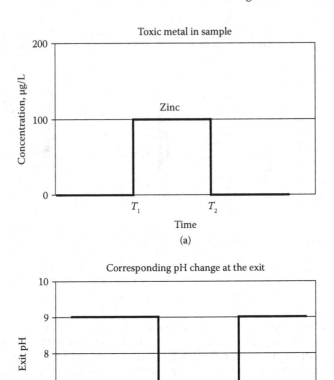

FIGURE 6.9 Schematic diagram illustrating sensing of toxic heavy metals in water through a change of pH using a conceptualized sorbent. (a) Step feed of toxic metal in the influent through the sorbent. (b) Resultant pH change at the exit of the sorbent after a lag period. (From Chatterjee, P. K., and SenGupta, A. K., *AIChE J*, 55 (11), 2997–3004, 2009; Chatterjee, P. K., *Sensing and Detection of Toxic Metals in Water with Innovative Sorption Based Techniques*, PhD Thesis, Civil and Environmental Engineering, Lehigh University, Bethlehem, PA, 2010.)

metals while maintaining a near-constant slightly alkaline pH, is conceived. The conceptualized sorbent, that is, HIM, mainly contains calcium magnesium silicate ($Ca_2MgSi_2O_7$) with a little or no amount of HFO. Calcium magnesium silicate is sparingly soluble in water that essentially maintains a near-constant alkaline pH by producing hydroxyl ions through slow hydrolysis according to the following reaction (the over bar indicates solid phase):

$$\overline{Ca_2MgSi_2O_7} + 3H_2O \leftrightarrow 2Ca^{2+} + Mg^{2+} + 2SiO_2 + 6OH^- \qquad (6.2)$$

Consequently, the pH of the surrounding solution always remains slightly alkaline (i.e., in the vicinity of 9.0), greatly enhancing the metal affinity of the surface hydroxyl groups in HIM. Thus, the HIM with the dual properties, that is, offering the selective metal binding surface sites while maintaining a steady alkaline pH in contact with water, embodies the attributes of the conceptualized sorbent illustrated in Figure 6.8d. However, sorbents synthesized from other combinations of materials such as aluminum oxide/zirconium oxide (site for metal binding) in combination with akermanite or gehlenite (maintaining alkaline pH through slow dissolution) are likely to exhibit the attributes of HIM.

6.3 SYNTHESIS AND CHARACTERIZATION OF CONCEPTUALIZED SORBENT

6.3.1 Synthesis of Hybrid Inorganic Sorbent Material (HIM)

The conceptualized sorbent material described previously must possess both selective binding sites with oxygen donor atoms for toxic metals and simultaneously maintain a near-constant alkaline pH in contact with aqueous solution. With reference from previous research investigations[45,46] a hybrid inorganic sorbent material (HIM) has been synthesized that shows high metal sorption affinity while maintaining an alkaline pH for a long time. HIM was synthesized in the laboratory using an operationally simple chemical-thermal technique.

6.3.2 Components of HIM

The active components of HIM are HFO and akermanite (a mineral with primary component calcium magnesium silicate). The HFO was prepared through a precipitation technique from a ferric chloride ($FeCl_3$) solution in deionized (DI) water (100 g/L) with the addition of 1 N sodium hydroxide solution (NaOH) until a pH ~11 was reached. The resultant ferric hydroxide $Fe(OH)_3$ suspension was allowed to settle for about 2 h before the supernatant was decanted. The precipitate thus obtained was rinsed with DI water for three cycles to ensure removal of excess sodium hydroxide and was further centrifuged for 15 min at 5000 rpm to get HFO. The freshly prepared HFO was then ground, air dried as shown in Figure 6.10a, and used for synthesis of HIM.

Akermanite is a mineral of sorosilicate $(Si_2O_4)^{6-}$ groups with major elements calcium, magnesium, silicon, and oxygen and named for Anders Richard Akerman, a Swedish metallurgist.[47] The akermanite sample used for synthesizing HIM is originally from Vesuvius, Italy, with the mineralogical data shown in Table 6.4. The chemical composition of the akermanite sample suggests primary components of calcium magnesium silicate with minor contributions from Al_2O_3 and FeO.[47–49] Figure 6.10b exhibits a photograph of a akermanite rock that is an off-white translucent crystal.[50] The akermanite sample was ground, washed with DI water, and air dried before using for HIM synthesis.

6.3.3 Steps for Synthesis of HIM

Freshly prepared crushed, air-dried HFO was thoroughly mixed with ground amorphous akermanite in the rotating batch receptacle for 24 h. The mixture was put

(a)   (b)

FIGURE 6.10 (a) Picture of crushed, dried, freshly prepared HFO. (b) Picture of Akermanite rock.

TABLE 6.4
Mineralogical Compositions and Salient Properties of Akermanite Rock

Primary Component	Calcium Magnesium Silicate	
Empirical formula	$Ca_2MgSi_2O_7$	
Composition	CaO	39.3%
	MgO	13.3%
	SiO_2	46.32%
	FeO	0.96%
	Al_2O_3	0.12%
Density	2.94 g/cm³	
Hardness (Mohs)	5–6	
Color	White with yellowish tone, translucent	
Crystal system	Tetragonal-scalenohedral	

Source: Swainson, I. P., et al., *Phys. Chem. Mineral, 19*, 185–195, 1992; mindat.org, *Akermanite*: Akermanite mineral information and data, http://www.mindat.org/min-70.html

---

under different levels of temperature for different time durations in a graphite furnace (Perkin-Elmer A-500) with a blanket of argon gas following drying, preheating, fusion, and cooling. In different batches of HIM synthesis, the mass ratios of HFO and akermanite were varied. The clinkers formed due to rapid fusion are subsequently ground, washed with DI water, and dried prior to use in the laboratory experiment. Figure 6.11a shows schematic illustrations, of the steps of the synthesis technique of HIM using HFO and akermanite, and Figure 6.11b shows an enlarged view of ground HIM particles used for further experiments.[38,50]

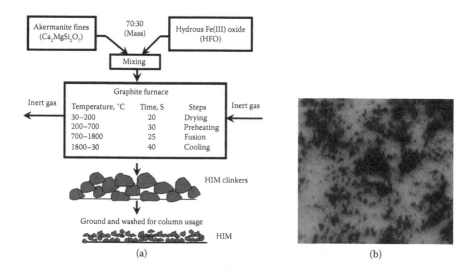

FIGURE 6.11   (a) Schematic illustration of steps for synthesis of HIM. (b) Picture of ground HIM particles (4× magnification). (From Chatterjee, P. K., and SenGupta, A. K., *AIChE J,* 55 (11), 2997–3004, 2009; Chatterjee, P. K., *Sensing and Detection of Toxic Metals in Water with Innovative Sorption Based Techniques,* PhD Thesis, Civil and Environmental Engineering, Lehigh University, Bethlehem, PA, 2010.)

6.3.4   Synthesis of **HIM-X**

The purpose of synthesizing HIM was to detect the presence of toxic heavy metals in water through a pH change. Upon promising experimental results with HIM, a material with reduced metal binding sites was prepared for the purpose of expediting the detection process. Leaving HFO, a well-recognized material for its ability of selective binding of heavy metals at slightly alkaline pH, hybrid inorganic material next generation (HIM-X) was synthesized using calcium oxide, magnesium oxide, and silica. Laboratory grade CaO, MgO, and SiO_2 at a molar ratio of 2:1:2 were thoroughly mixed before putting into the graphite furnace for rapid thermal fusion with the following steps[50,51] as illustrated in Figure 6.12a:

- Drying for 15 s (30–200°C)
- Preheating 15 s (200–700°C)
- Ramp up 5 s (700–2000°C) and fusion 20 s (2000°C)
- Cool down time 20 s (2000–50°C)

Clinkers of irregular shapes produced from the graphite furnace were crushed and made ready for further experimentation. Figure 6.12b shows a picture of crushed HIM-X particles.

6.3.5   Analysis

Samples were analyzed for lead, zinc, copper, nickel, and so on, using a Perkin-Elmer AAS with a graphite furnace accessory (model SIMA 6000) following the

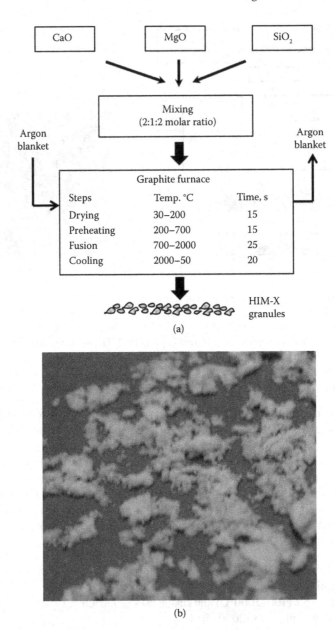

(a)

(b)

FIGURE 6.12   (a) Schematic representation of steps of synthesis of HIM-X. (b) Picture of ground HIM-X particles used for experimentation (3× magnification). (From Chatterjee, P. K., *Sensing and Detection of Toxic Metals in Water with Innovative Sorption Based Techniques*, PhD Thesis, Civil and Environmental Engineering, Lehigh University, Bethlehem, PA, 2010; Chatterjee, P. K., and SenGupta, A. K., Toxic metal sensing through novel use of hybrid inorganic and polymeric ion, *Solvent Extraction Ion Exchange*, 29, 398–420, 2011.)

protocol of standard methods.[52] Calcium was analyzed using an atomic absorption spectrophotometer with a flame attachment (Perkin-Elmer model: Analyst 200). Sodium, chloride, and sulfate were analyzed using a Dionex Ion Chromatograph (model DX-120 IC). Inorganic carbon and total organic carbon (TOC) were analyzed using a SHIMADZU carbon analyzer. The sample pH was recorded using an Accumet meter (XL 15). SEM and EDX analyses were performed using a Hitachi scanning electron microscope (SEM-EDX, model no. 4300). The BET surface area was measured using MICROMERITICS ASAP 2020. The zeta potential for HIM particles was analyzed using an Anton Paar Electrokinetic Analyzer (model: SURPASS) using the laser Doppler electrophoresis technique.

6.3.6   CHARACTERIZATION

6.3.6.1   Physical Properties: Size, Porosity, and Density

The physical characteristics of HIM and HIM-X are listed in Table 6.5. Both the products are composite granular materials containing a matrix of different metal oxides. Analyses of different parameters such as void volume, pore volume, bulk, and true densities clearly suggest significant porous configurations. Particle size analysis is done using a Beckman-Coulter particle counter (model: LS 100Q), which results in average mean diameters of 372 and 225 μm for HIM and HIM-X, respectively.

6.3.6.2   Surface Area

Surface area is one of the important parameters that determine the sorption ability of a sorbent. Surface area for inorganic sorbents such as different types of iron oxides, alumina, and silicates are reported in the literature though figures are found to vary over a wide range depending on the method of measurement, particle size, porosity,

---

TABLE 6.5
Physical Characteristics of HIM and HIM-X

Parameters	HIM	HIM-X
Mean particle size (projected diameter)	372 μm with SD = 187 μm	225 μm with SD = 107.5 μm
Pore volume	0.2962 mL/g	0.384 mL/g
Void volume	41.2%	44.8%
Bulk density	1.52 g/mL	1.38 g/mL
True density	3.5 g/mL	2.9 g/mL
Average BET surface area	11.7 m²/g	20.4 m²/g

Source: Chatterjee, P. K., *Sensing and Detection of Toxic Metals in Water with Innovative Sorption Based Techniques,* PhD Thesis, Civil and Environmental Engineering, Lehigh University, Bethlehem, PA, 2010.

and crystalline structure. The specific surface areas of iron oxides are reported to vary from 150 to 300 m²/g[29] and activated alumina with approximate surface area ~ 10 m²/g.[53]

The Brunauer–Emmett–Teller (BET) surface area measurements for HIM and HIM-X were performed using a nitrogen adsorption experiment. The adsorption multi-point data were obtained at different pressures relative to 760 mmHg. Figure 6.13a and b shows nitrogen (N_2) sorption/desorption isotherms for HIM and HIM-X, respectively, through a plot of adsorption quantity (Q) versus relative pressure (P/P_o). Further, based on the sorption data, a resultant linear fit for the relative pressure range $0.05 < P/P_o < 0.35$ is obtained from the following BET equation[54]:

$$\frac{1}{Q\left[\left(\frac{P_o}{P}\right)-1\right]} = \frac{C-1}{Q_mC}\left(\frac{P}{P_o}\right) + \frac{1}{Q_mC} \tag{6.3}$$

with Q being the absorbed gas quantity (cm³/g) and P/P_o the relative pressure. A linear plot of $1/Q[(P_o/P) - 1]$ versus (P/P_o) allows determination of Q_m (monolayer absorbed gas quantity) and C (BET constant) from the slope and intercept of the linear plot. The specific BET surface area is further obtained by

$S = Q_m$ * Avogadro's No.* area of one N_2 molecule/molar volume of N_2

The BET surface area for HIM and HIM-X thus obtained is 11.7 and 20.4 m²/g, respectively.[50]

6.3.6.3 Scanning Electron Microscopy

SEM analysis images of the sample surface were obtained by scanning it with a high energy beam of electrons. It provides useful information about the surface topography, external morphology (texture), porosity, pore geometry, composition, and so on. SEM photographs of HIM particles demonstrated in Figure 6.14a and b at different magnifications suggest that each particle is an aggregate of several micro-particles containing large macropores of irregular shapes and sizes. Figure 6.15a and b exhibits SEM images for HIM-X at two different magnifications. The images in Figures 6.14b and 6.15b suggest that each particle consists of a network of pores with macroparticles as an agglomerate of several microparticles containing pores. It is observed that the microparticles that constitute HIM-X seem to be much smaller (50–400 nm) than the microparticles found in HIM (100–700 nm). This observation is consistent with particle size analysis using a Coulter particle counter for the two samples.

Energy dispersive X-ray analysis commonly known as EDX or EDAX is applied in conjunction with SEM analysis to identify the elemental composition of a sample material or a small area of interest in the sample. EDX spectroscopy of the surface provides qualitative information on different elements that are present.[55] Such analyses have been performed for both fresh and used samples of HIM and HIM-X, and are elaborated later in this chapter.

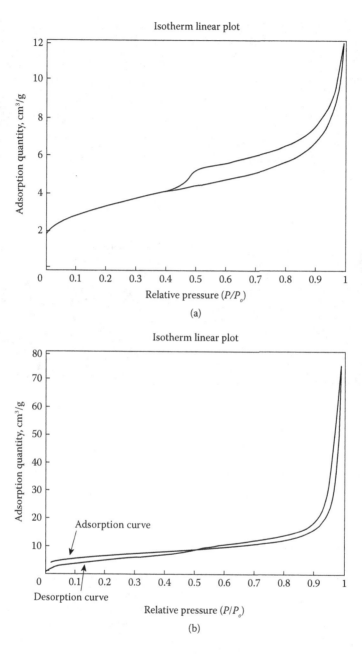

FIGURE 6.13   Nitrogen (N_2) sorption–desorption isotherm plots related to BET surface area measurement for (a) HIM particles and (b) HIM-X particles. (From Chatterjee, P. K., *Sensing and Detection of Toxic Metals in Water with Innovative Sorption Based Techniques*, PhD Thesis, Civil and Environmental Engineering, Lehigh University, Bethlehem, PA, 2010.)

FIGURE 6.14 Scanning electron microscopy (SEM) for an HIM particle. (a) At (200×) magnification and (b) at (3000×) magnification.

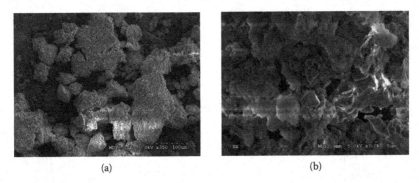

FIGURE 6.15 Scanning electron microscopy (SEM) for HIM-X particle. (a) At (350×) magnification and (b) at (7000×) magnification.

6.4 EXPERIMENTAL VALIDATION: METAL SENSING THROUGH pH SWING

6.4.1 FIXED-BED COLUMN RUN

Fixed-bed column runs were carried out using HIM or HIM-X bed in a mini-glass column of 7 mm diameter, a constant flow pump (Fluid Metering, Inc.), and a fraction collector (ELDEX). For different column runs, about 150–200 mg of sorbent material was used. The empty bed contact time (EBCT), 9–12 s, was maintained and the superficial liquid velocity (SLV) was recorded for each run. Other hydrodynamic parameters and characteristics of the column run were similar to those described elsewhere.[56,57] For every run, to ensure the dissolved state of toxic metals in feed solution, the concentration was usually kept below the solubility limit.

6.4.2 STEADY ALKALINE pH IN CONTACT WITH WATER

The ability of HIM and HIM-X to maintain a steady alkaline pH in contact with water for a prolonged time was confirmed through dissolution studies. Two separate batch dissolution studies with HIM and HIM-X in DI water (~18.2 MΩ) and 0.005 M

NaCl solution were conducted. The material (HIM or HIM-X) of 0.1 g was added to 100 mL DI water or 0.005 M NaCl solution in a bottle, stirred for 4 h, and the pH was recorded. Subsequently, the slurry was filtered and the recovered material was then added to a fresh batch of 100 mL solution. Altogether three such cycles were carried out; Figure 6.16a and b shows the pH versus time plot for all three cycles of the dissolution studies of HIM and HIM-X. Two important things are noted. Both HIM and HIM-X quickly reach a pH above 9.6 even after the third cycle. There is no noticeable variation of equilibrium pH between DI water and 0.005 M NaCl solution.[50]

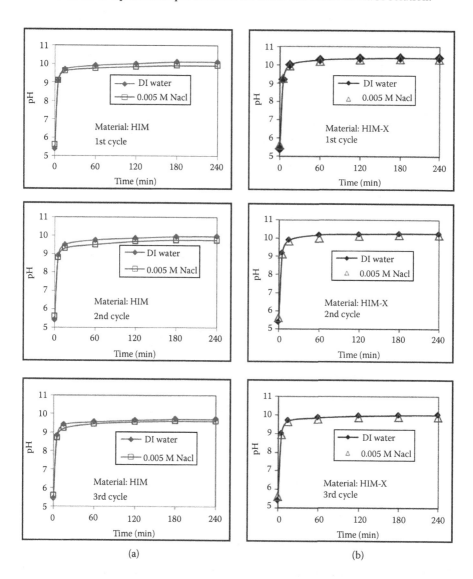

(a)   (b)

FIGURE 6.16 Solution pH versus time plot for three consecutive cycles of batch hydrolysis tests with DI water and with dilute NaCl solution. (a) HIM and (b) HIM-X.

Further, fixed-bed column experiments containing HIM and HIM-X were carried out separately with an influent aqueous solution containing commonly encountered ions (e.g., Na^+, Ca^{2+}, Cl^-, SO_4^{2-}) for an EBCT 9–10 s. Figure 6.17a and b shows that in the presence of commonly encountered electrolytes, the pH at the exit of the column always remained slightly higher than 9.0 due to hydrolysis of HIM and HIM-X. For comparison, when the influent solution was replaced by DI water (~18.2 MΩ), the pH at the column exit in both the cases remained around 9.0, showing practically no significant fluctuation.[38,50,51]

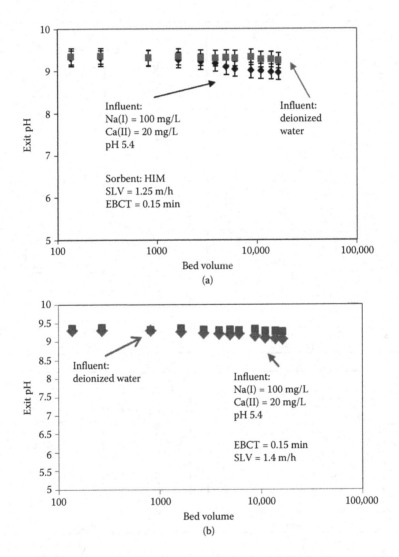

FIGURE 6.17 Plot of exit pH versus bed volume (BV) for fixed-bed column run with feed sample containing common electrolytes and deionized water. (a) Bed containing HIM and (b) bed containing HIM-X.

6.4.3 DROP IN pH IN THE PRESENCE OF TOXIC METALS

6.4.3.1 Change of pH for Trace Zinc in the Influent

It is already validated that if an aqueous solution containing only common electrolytes (no toxic metal) is passed through a bed of HIM, the column effluent pH remains alkaline. However, when the trace concentration of zinc (0.5 mg/L) is added to the same feed, a sharp drop in pH accompanied by a breakthrough of zinc in the effluent is observed. Figure 6.18a demonstrates how the exit pH from the mini-column sharply dropped (from 9 to <7) after 6500 bed volumes (BVs). The drop in the pH coincided with the breakthrough of zinc at the effluent. The change in the pH was validated by a phenolphthalein test showing the pink color for alkaline pH (no zinc) and turning colorless with the reduction of pH (zinc gradually broke through) as shown in Figure 6.18b. After approximately 12,000 BVs, the influent was replaced by feed containing no zinc. Subsequently, the column exit pH rose to the alkaline domain (around 9.0) again, the exit zinc concentration dropped, and the solution turned pink with an addition of a drop of phenolphthalein indicator (Figure 6.18a and b). This observation strongly suggests that the pH at the exit of the column is responsive to the fluctuation of zinc concentration in the feed solution. A similar experiment with HIM-X bed for influent containing trace zinc (0.5 mg/L) along with other common electrolytes resulted in sharp pH drop (from 9 to <7) accompanied

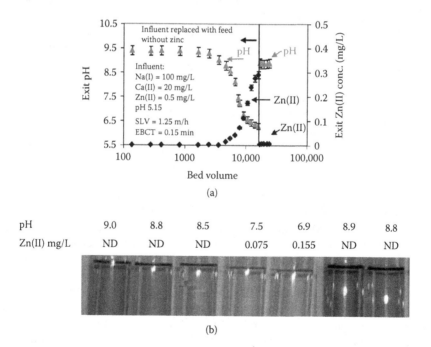

(a)

pH	9.0	8.8	8.5	7.5	6.9	8.9	8.8
Zn(II) mg/L	ND	ND	ND	0.075	0.155	ND	ND

(b)

FIGURE 6.18 (a) Plot of exit pH and zinc concentration versus bed volume for HIM column run. (b) Visual change in color of phenolphthalein indicator for exit samples from pink to colorless with zinc breakthrough and again to pink upon withdrawal of zinc from the influent (ND for "not detectable").

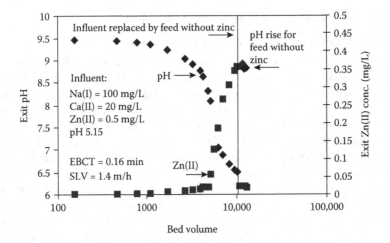

FIGURE 6.19 Plot of exit pH and zinc concentration versus bed volume for HIM-X column run with feed solution containing Zn (0.5 mg/L) along with other common background electrolytes.

with zinc breakthrough as shown in Figure 6.19, producing an identical result that obtained for experiment with HIM (Figure 6.18a and b). However, in case of HIM-X, the pH dropped earlier after 5000 BVs approximately in comparison to 6500 BVs for HIM for the identical experiment with trace zinc. After around 10,000 BVs, when the feed was replaced by influent containing no zinc, the exit pH quickly rose to the alkaline zone around 9.0.[50]

6.4.3.2 Development of Characteristic Peak from Slope of pH Profiles for Trace Zinc

The sharp change in the exit pH of the HIM column in the presence of trace zinc (0.5 mg/L) shown in Figure 6.18a can be represented through a meaningful mathematical expression. The negative derivative of the exit pH profile produces a steep peak with respect to a sudden drop in the pH, as commonly observed in chromatographic analytical instruments. Quantitatively, the slope of the pH breakthrough curve (in other words, $-dpH/dBV$) exhibits a distinctive peak. Such a peak occurs because the negative slope of the pH breakthrough curve (i.e., $-dpH/dBV$) gradually increases, goes through a maximum, and then drops. Figure 6.20 shows the plot of ($-dpH/dBV$) versus BV for the data presented in Figure 6.18a for 0.5 mg/L of zinc in the feed. The HIM column run for influent containing 1 mg/L of Zn(II) with otherwise identical background and experimental conditions produces a sharp drop in exit pH about 3500 BVs approximately as shown in Figure 6.21a. The derivative of the pH plot in Figure 6.21a results in a meaningful peak in Figure 6.21b. A comparative evaluation of peaks between Figures 6.20 and 6.21b clearly suggests that peak position and height are strongly dependent on feed Zn(II) concentration. For 0.5 mg/L zinc in the feed, a peak appeared after 6500 BVs with a lower height in comparison with the peak for 1 mg/L Zn in the feed which showed up much earlier after 3500 BVs.[38,50]

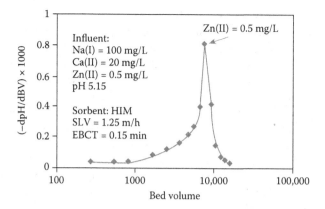

FIGURE 6.20   Plot of (−dpH/dBV) versus bed volume showing peak for trace zinc (0.5 mg/L) in the influent based on the data presented in Figure 6.18a for column run using HIM.

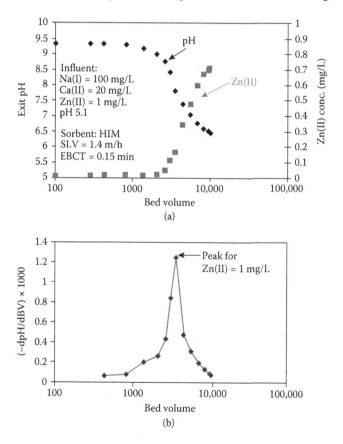

FIGURE 6.21   (a) Effluent pH and zinc profile for HIM column run with feed containing 1 mg/L zinc with the background of usual electrolytes. (b) Negative derivative of pH (−dpH/dBV) versus bed volume plot for data in (a).

6.4.3.3   Characteristic Peaks for Other Toxic Metals

A similar significant pH drop in the presence of trace concentration of other toxic metals such as Ni, Cu, and Pb at the exit of column with HIM was observed. Figure 6.22a, b, and c represents peaks (–dpH/dBV vs BV plot) for three separate column runs with HIM for influent containing 0.5 mg/L Ni(II), Cu(II), and Pb(II), respectively, with the usual electrolyte background under identical hydrodynamic conditions and bed configurations. Here also, a drop in pH coincided with the breakthrough of the particular heavy metal. A comparison among characteristic peaks for Zn (Figure 6.20), Ni, Cu, and Pb in the feed sample at the same mass concentration (0.5 mg/L) reveals that as a single toxic metal in the feed peaks of Ni and Zn appeared close to each other with comparable heights, the peaks for Cu and Pb emerged much later with larger peak heights. This basically suggests that Pb (molecular weight = 207) has a lower molar concentration than Ni, Zn, and Cu for the same mass concentration, and Cu has much higher sorption affinity than Zn and Ni toward the HIM metal-binding sites. Thus, qualitatively characteristic peaks depend on both the type and concentration of the toxic metals. In other words, under identical conditions the presence of a single toxic metal in the influent would result in a drop in pH differing in magnitude and occurring at different times.[38,51]

6.4.3.4   Test for Trace Lead in Feed Sample

Figure 6.23a and b shows the pH profiles and peaks (–dpH/dBV plot) for trace lead concentrations at 75 and 55 µg/L, respectively, for two separate column runs with HIM-X under identical experimental conditions and bed parameters. Note that for both the concentrations, the pH dropped sharply from near 9.0 to about 7.0 at around 10,000 and slightly after 10,000 BVs, respectively. Peaks emerged with distinctive heights. Comparing with peaks of two different lead concentrations (75 and 55 µg/L) two important points were noted: first, the peak of 55 µg/L came after that for 75 µg/L; second, the peak for 75 µg/L is greater than that for 55 µg/L.[38,50] Thus, in addition to sensing the presence of a trace toxic metal in water, semi-quantitative detections might be possible using the present technique.

6.4.3.5   Presence of More Than One Toxic Metal

A feed sample containing both zinc and copper, when passed through a column of HIM-X, resulted in a pH profile and peak as explained in Figure 6.24a and b. When present as a single metal, the peak of copper appeared much after zinc as observed in Figures 6.20 and 6.22a, b, and c following an affinity sequence of $Pb^{2+} > Cu^{2+} \gg Zn^{2+} \geq Ni^{2+}$ for the metal-binding sites of the sorbent. A similar affinity sequence for binding onto iron oxide and/or metal oxide surfaces had been validated by many previous investigations.[28,29,53,58] However, in this case the peak of zinc preceded the copper peak with a relatively shorter peak height for copper in comparison with the peak observed when copper alone was present at the same concentration. The exit pH profile and metal concentration for test results with feed containing multiple toxic metals such as Zn, Ni, Cu, and Pb are demonstrated in Figure 6.25. The resultant peaks obtained from the negative derivative (–dpH/dBV) of slope of pH curve could not be attributed to a particular toxic metal. However, a sharp drop in pH in the presence of either single or multiple toxic metals clearly signals the presence of toxic metal(s)

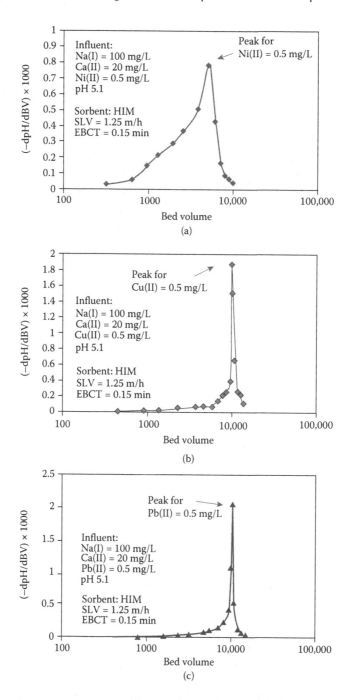

FIGURE 6.22 Negative slope of pH profile (−dpH/dBV) versus bed volume plot for column run with HIM for feed solution containing trace concentration of different toxic metals: (a) nickel, (b) copper, and (c) lead, each with usual background electrolytes.

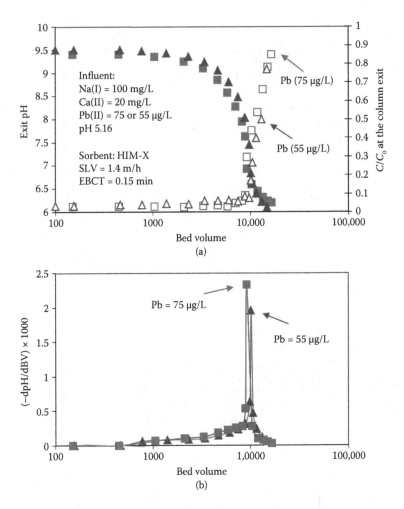

FIGURE 6.23 (a) Effluent pH and Pb profile for two separate HIM-X column runs with feed containing trace concentrations of lead. (b) Peaks of lead from ($-dpH/dBV$) versus bed volume plot corresponding to data of (a).

and may well be used for diagnosing the presence of toxic metals. For simultaneous detection of two toxic metals, different peak times (elution times) may be useful for metals with a large difference in affinity for the sorption sites (e.g., zinc and copper in Figure 6.24a and b). However, multiple toxic metals with close sorption affinities present in a sample though result in a distinctive pH drop signaling the presence of toxic metals; identification of respective metals from their peak height cannot be achieved.

6.4.3.6 Detection of Ultra-Low Concentration of Toxic Metals in Water

In order to detect very low concentrations (usually 10 µg/L or less) of target metals, a simple preconcentration technique was applied to enhance the sensitivity of the present method. A typical synthesized sample of 2500 mL spiked with lead of less than

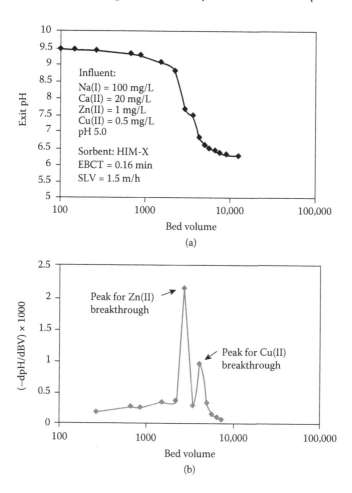

FIGURE 6.24 (a) Effluent pH for fixed-bed run through HIM-X with sample containing both zinc and copper. (b) (–dpH/dBV) versus bed volume plot showing peaks for copper and zinc corresponding to run in (a).

10 µg/L [Pb(II)] was passed through a column containing a chelating ion exchanger (Chelex-100) and was subsequently eluted by 250 mL, 1% sulfuric acid solution which resulted in about 10 times the preconcentration of lead. Similar evidence of a preconcentration technique applying sorption–desorption of a chelating ion exchanger are available in the open literature.[59,60] After necessary pH adjustment, the preconcentrated regenerant was passed through a bed of HIM-X in a tiny column (4 mm diameter) for detection of lead through pH change. Figure 6.26 exhibits the plots of (–dpH/dBV) versus bed volume for both parent and preconcentrated (10×) samples. The preconcentrated sample shows a distinctively sharp peak compared to the parent sample. The lead concentration (78 µg/L) in the preconcentrated sample was almost 10 times that of the concentration of lead in the parent sample (8 µg/L), which basically confirms the precision of this preconcentration method.

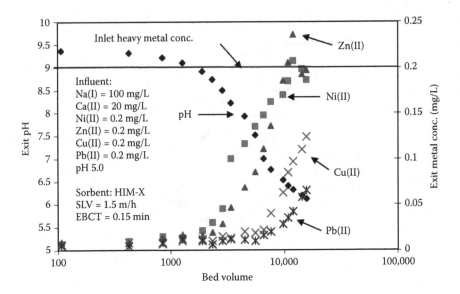

FIGURE 6.25 Effluent pH and metal concentrations for feed containing multiple toxic metals through HIM-X bed.

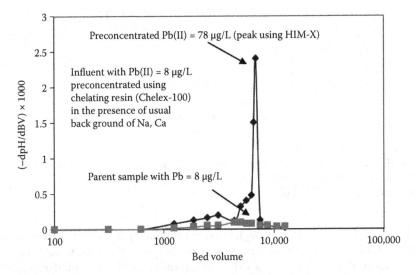

FIGURE 6.26 Plot of (−dpH/dBV) versus bed volume for demonstration of detection of ultra-low concentration of lead (8 µg/L) by preconcentration technique through HIM-X bed.

6.5 OVERCOMING INTERFERENCE OF BUFFERS

Since the pH is the measuring parameter, the presence of any buffering solutes (e.g., carbonate, phosphate, or other weak-acid anions) in the feed is likely to interfere with the detection process by influencing the change in pH. Within the characteristic pH range of 6.5 to 8.5 of most natural waters, bicarbonate (also carbonate) provides the resistance to change in pH. In addition, different water sources may contain common weak-acid anionic ligands such as phosphates and NOM (which is mainly composed of weak-acid anions such as fulvate and humate) that act as buffering agents.[2,11] Since pH is a surrogate indicator for the presence of toxic heavy metals, any buffering solute present in the feed along with toxic metal is likely to interfere with this detection technique.

6.5.1 Tap Water Spiked with Trace Lead—Test for Influence of Carbonate System

The carbonate system controls the pH of most natural waters. Atmospheric carbon dioxide (CO_2) maintains a natural buffer through speciation of H_2CO_3, HCO_3^-, and CO_3^- at different pHs. The commonly known useful relationships[11,37] of these species are

$$[HCO_3^-] / [H_2CO_3] = 10^{pH-6.35}$$

and

$$[HCO_3^-] / [CO_3^{2-}] = 10^{10.33-pH}$$

Within the characteristic pH range 6.5 to 8.5 of most natural water, bicarbonate (HCO_3^-) is a predominant carbonate species that is likely to interfere with change in the pH for sensing of toxic metals. In order to overcome interference of carbonate species, the influent pH was adjusted to less than 5.0 with simultaneous purging of N_2 gas that stripped off carbonate from aqueous solution in the form of carbon dioxide. Figure 6.27 exhibits ($-dpH/dBV$) versus bed volume plots for two identical

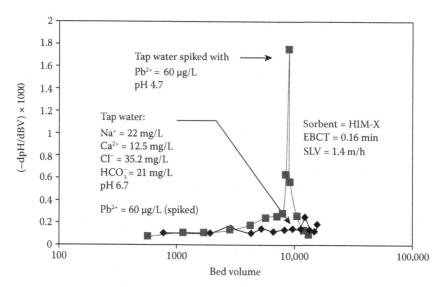

FIGURE 6.27 Plot of ($-dpH/dBV$) versus bed volume for Bethlehem city tap water spiked with trace lead showing interference of carbonate species and subsequent result for overcoming interferences of carbonate species.

HIM-X column runs using influent of Bethlehem city tap water duly spiked with 60 µg/L lead with and without adjustment of the influent pH. A run with pH adjustment (i.e., pH 4.7) produced a sharp, distinguishable peak for trace lead in contrast to the sample without pH adjustment (i.e., the pH same as tap water ~6.7), which produced a diffused, non-distinguishable peak, confirming the influence of bicarbonate toward the resistance in pH change.

6.5.2 INTERFERENCE OF PHOSPHATE, NOM, AND SIMILAR ANIONIC LIGAND

Phosphate is often added as a corrosion inhibitor in municipal and industrial water systems. Run-off from agricultural field dosed with phosphate-based fertilizer can be another important source of phosphate. Phosphorous occurs in soil and different rocks and geochemical leaching may also contribute to ingress of phosphate in water.[61,62] Phosphate is a triprotic (H_3PO_4) system, producing non-volatile buffering species that are likely to interfere with the sensing technique of toxic metals through pH change. In a similar vein, NOM, which essentially comprises weak aliphatic and aromatic ligands (e.g., fulvate), is also present in rivers, lakes, and surface water supplies and generally in anionic forms under the prevailing pH conditions of natural water.[38,63,64] These weak-acid anions (Lewis bases) may sequester toxic metal cations (Lewis acids) through complex formation, and they are also likely to resist a change in pH and interfere with the sensing process. To overcome the interference of phosphate, NOM, or similar anionic ligands, a modification of the bed has been conceived so as to arrest phosphates/NOM before the heavy-metal laden solution reaches the HIM-X bed.

6.5.2.1 Selective Removal of Phosphate and Other Anionic Ligands through HAIX

Previous studies revealed that ferric oxide nanoparticles loaded with hybrid anion exchange resin (HAIX) can selectively uptake phosphate, arsenate, oxalate, or similar anionic ligands in preference to common anions like chloride, sulfate, and so on, from water following Donnan membrane equilibrium.

In the present study, HAIX prepared from a macroporous strong base anion exchange resin (Purolite A-400 or equivalent) loaded with HFO nanoparticles in the exchanger phase has been used. A step-wise protocol for HAIX preparation is already available in the open literature[65–67] and not repeated here. The average particle size of HAIX beads varies 400–800 µm with mean of ~500 µm.[66,67] Figure 6.28a displays an enlarged view (40×) of HAIX beads. Note that the spherical geometry of the original resin particles is retained after processing for loading of HFO nanoparticles inside the resin phase. It is worth noting that the hybrid particles develop a reddish brown color after loading of HFO. Energy dispersive X-ray spectroscopy (EDX) analysis of iron count along the diameter of a sliced HAIX bead reveals the pattern of HFO loading displayed in Figure 6.28b. It explains higher iron content near the edge of the bead and gradually decreases as it proceeds toward the center, which essentially reveals a greater concentration of HFO nanoparticles near the periphery compared to that at the core of the bead. In HAIX, nanoscale HFO particles are irreversibly dispersed within the polymeric

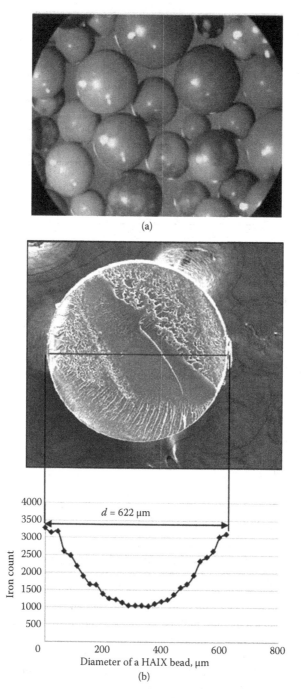

FIGURE 6.28 (a) Enlarged views of hybrid anion exchanger (HAIX) spherical particles. (b) Energy dispersive X-ray mapping of iron along the diameter of a sliced spherical HAIX particle (SEM image 120×).

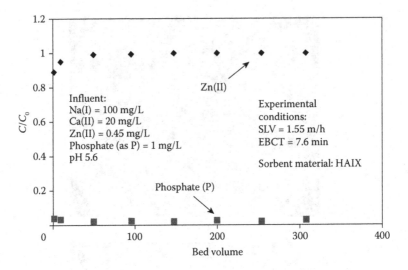

FIGURE 6.29 Concentration profiles of zinc and phosphate at the exit of a column containing HAIX media.

phase of anion exchanger. A high concentration of fixed positively charged quaternary ammonium functional groups (R_4N^+) in the exchanger phase results in a Donnan membrane effect that greatly enhances the ligand sorption capacity of HFO particles within the anion exchanger. Figure 6.29 demonstrates the result of a fixed-bed column run using HAIX media with the feedwater containing trace phosphate and zinc with a background of high concentration of commonly occurring electrolytes $(Na^+, Ca^{2+}, Cl^-, SO_4^{2-})$. At the column exit, phosphate is completely removed as opposed to zinc which breaks through instantaneously, showing complete rejection. This observation is in compliance with the results reported in previous investigations.[61,62,66–69]

6.5.2.2 Toxic Metal Sensing Overcoming Interference of Phosphate and NOM

The interfering effects of phosphate and NOM are overcome by putting HAIX ahead of HIM-X in the column so that the feed sample passes through HAIX first, thus only the phosphate (or other ligand) free sample comes in contact with HIM-X. Figure 6.30a shows the peaks for trace zinc (0.48 mg/L) when phosphate is also present in the sample. The result of experiment in a column containing HAIX ahead of HIM-X generates a greater and distinct peak as compared to a flat and diffused peak obtained for HIM-X alone. In the case of a control solution that has identical components without zinc or other toxic metals, a flat peak is observed confirming that the peak is attributed to the presence of zinc in the feed solution. Figure 6.30b demonstrates the results in the presence of NOM. Lehigh River water containing TOC 12 mg/L duly spiked with 0.5 mg/L zinc produced a very sharp peak for the combined HAIX and HIM-X column. In comparison, the peak with HIM-X alone was shorter and more diffused in the presence of NOM.

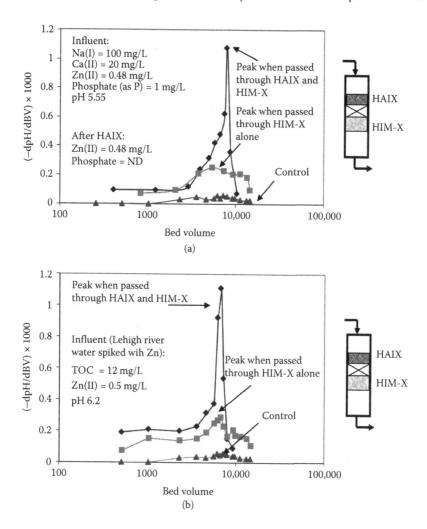

FIGURE 6.30 (a) Plot of (−dpH/dBV) versus bed volume showing interference phosphate and overcoming interference using HAIX prior to HIM-X in a column for trace zinc (0.48 mg/L). (b) Plot of (−dpH/dBV) versus bed volume showing interference NOM and overcoming interference using HAIX prior to HIM-X in a column for trace zinc (0.5 mg/L).

Phosphate or another anionic ligand is selectively picked up by HAIX in preference to other common anions following Donnan membrane equilibrium while Zn^{2+} is completely rejected according to the Donnan co-ion exclusion principle as observed in Figure 6.29.

6.5.2.3 Donnan Membrane Equilibrium and Co-Ion Exclusion Effect

The Gibbs–Donnan equation that describes the "Donnan membrane equilibrium" arises from the unequal distribution of the mobile ions[70] or, in other words, the inability of ions to diffuse from one phase to the other through interface. The polymeric phase of an ion exchanger can be viewed as a polyelectrolyte where functional groups (e.g., quaternary ammonium or R_4N^+ groups for anion exchangers and sulfonic acid or SO_3^- groups for cation exchangers) are covalently attached and, hence, non-diffusible. Both counterions and co-ions in the bulk solution phase are mobile and can move freely under chemical or electrical potential gradient. The presence of fixed functional groups in the exchanger phase makes the ion exchanger a semi-permeable membrane that essentially gives rise to the development of the Donnan potential.[70,71]

According to the Donnan principle of ion exchange equilibria, the electro-chemical potential of all ionic species (regardless of whether they are counter- or co-ions) in the ion exchanger phase will be equal to that in the bulk liquid phase. The electrochemical potential of any ion is related to its activity and valence and is expressed by the following relation[70–72];

Thus accordingly,

$$\left(\frac{\overline{a}_1}{a_1}\right)^{1/z_1} = \left(\frac{\overline{a}_2}{a_2}\right)^{1/z_2} = \cdots \left(\frac{\overline{a}_n}{a_n}\right)^{1/z_n} \tag{6.4}$$

with a and z referring to the activity and valence of the ions, respectively. The sub-scripts denote specific ions and the overbar refers to the ion exchange or solid phase. To illustrate the Donnan membrane effect, let us consider an HAIX bead in which HFO nanoparticles are dispersed within the exchanger phase of the strong base anion exchange resin and are immersed in a large volume of solution containing 10 mM (0.01 M) NaCl and 10^{-2} mM (10^{-5} M) $ZnHPO_4$ as explained in Figure 6.31a. In the solution, the concentrations of dissolved Zn^{2+} and HPO_4^{2-} are negligible compared to the concentrations of Na^+ and Cl^-. The HAIX bead with a non-diffusible quaternary ammonium group (R_4N^+) in the exchanger phase is originally in chloride form and the exchange capacity is 1.0 M. According to the electroneutrality condition in the exchanger phase of original HAIX bead,

$$[R_4N^+]_R = [Cl^-]_R = 1.0 \text{ M} \tag{6.5}$$

The subscript "R" denotes the resin phase. All cations and anions except non-diffusible R_4N^+ will redistribute between exchanger and aqueous phase following the Donnan equilibrium principle. Assuming ideality, the activity term in the Donnan equilibrium relationship (Equation 6.4) may be replaced by molar concentrations, and the equilibrium relationship is derived by the following equation (subscripts R and W refer to resin and water phase, respectively):

$$\frac{\left[Na^+\right]_R^2}{\left[Na^+\right]_W^2} = \frac{\left[Cl^-\right]_W^2}{\left[Cl^-\right]_R^2} = \frac{[Zn^{2+}]_R}{[Zn^{2+}]_W} = \frac{[HPO_4^{2-}]_W}{[HPO_4^{2-}]_R} \tag{6.6}$$

The resin phase electroneutrality is

$$[R_4N^+]_R + 2[Zn^{2+}]_R + [Na^+]_R = 2[HPO_4^{2-}]_R + [Cl^-]_R \tag{6.7}$$

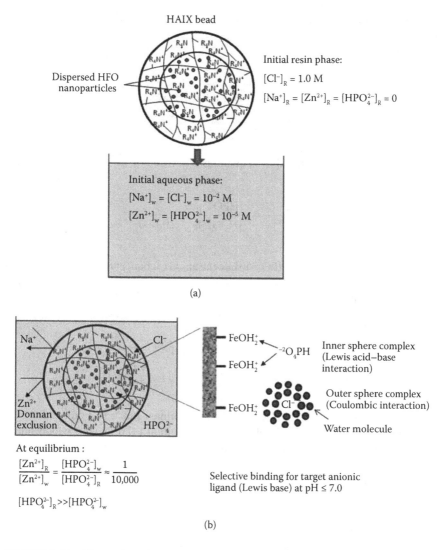

FIGURE 6.31 Illustration of equilibrium principle for a HAIX bead with irreversibly dispersed HFO particles within exchanger phase in contact with solution containing trace zinc and phosphate along with background electrolyte (NaCl). (a) Initial condition before immersion of HAIX bead. (b) At equilibrium after immersion of HAIX bead in solution.

The aqueous phase electroneutrality is

$$2[Zn^{2+}]_W + [Na^+]_W = [Cl^-]_W + 2[HPO_4^{2-}]_W \qquad (6.8)$$

For the experimental conditions in the aqueous phase, $[Na^+]_W \gg [Zn^{2+}]_W$ and $[Cl^-]_W \gg [HPO_4^{2-}]_W$. So, from Equation 6.8 it may be written

$$[Na^+]_W = [Cl^-]_W = 0.01 \text{ M}$$

Following Equation 6.4, $[Na^+]_R [Cl^-]_R = [Na+]_W [Cl^-]_W$, with $[Cl^-]_R = 1$, $[Na^+]_R = (0.01)^2 = 0.0001$ M.

Again from Equation 6.6,

$$\frac{\left[Na^+\right]_R^2}{\left[Na^+\right]_W^2} = \frac{[Zn^{2+}]_R}{[Zn^{2+}]_W} = \frac{[HPO_4^{2-}]_W}{[HPO_4^{2-}]_R} \qquad (6.9)$$

Thus, simplification of the above equations will result in

$$\frac{[Zn^{2+}]_R}{[Zn^{2+}]_W} = \frac{[HPO_4^{2-}]_W}{[HPO_4^{2-}]_R} \approx \frac{1}{10,000}$$

Thus, the concentration of HPO_4^{2-} inside the anion exchanger resin phase is several orders of magnitude greater than HPO_4^{2-} in the aqueous phase. HFO nanoparticles are dispersed within the resin phase of HAIX. The protonation and de-protonation reactions of surface functionalities of HFO particles[29] are represented as follows:

$$FeOH_2^+ \quad \leftrightarrow \quad H^+ + FeOH \quad pKa_1 = 6.6$$

$$FeOH \quad \leftrightarrow \quad H^+ + FeO^- \quad pKa_2 = 8.8$$

Thus, HFO particles inside HAIX offers predominant functional groups $FeOH_2^+$ and FeOH at the prevailing pH of the sample solution (~7.0 or less) which binds phosphate or similar anionic ligands through Lewis acid–base interactions in preference to common anions such as sulfate and chloride as illustrated in Figure 6.31b. Note that HAIX beads are electrically neutral and do not influence the permeation of phosphate into the polymeric phase of a hybrid ion exchanger. The permeability of phosphate in the resin phase is greatly enhanced due to the presence of high concentrations of non-diffusible fixed positive charges (R_4N^+) in the polymer phase of exchanger.[71,72] The experimental result illustrated in Figure 6.29 also validates selective removal of trace phosphate from the background of high concentration of common anions (Cl$^-$, SO$_4^{2-}$). Thus, selectivity of phosphate, NOM, or another anionic ligand can be greatly enhanced in respect to common anions following Donnan equilibrium. In contrast, Zn^{2+} in the aqueous phase is several orders of magnitude greater, resulting in almost complete rejection according to the Donnan co-ion exclusion effect[51,66,68,71] as is evident from the experimental result shown in Figures 6.30a and b.

The sample containing interfering phosphate/NOM is allowed to pass through a bed of specially prepared hybrid anion exchanger (HAIX) ahead of the HIM-X sorbent bed. Thus, the solution exiting from HAIX is free from phosphate (or fulvate, etc., in NOM) but the concentration of zinc remains unchanged. Passing through the HIM-X bed gives a subsequent pH drop and distinctive peaks are obtained as illustrated in Figure 6.30a and b.

6.6   CHALLENGES TOWARD REDUCTION OF DETECTION TIME AND SAMPLE VOLUME

Use of HIM-X though decreased the time requirement for sensing of toxic metals through change of pH with respect to HIM as explained in earlier section, however, for all practical applications further improvement in reduction of detection time (and the sample volume as well) is necessary. About 700 mL of sample water is needed to pass through the HIM-X standard column (7 mm diameter) to observe pH drop for 0.5 mg/L zinc in a HIM-X column. The requirement of sample volume to pass through may further increase for lead and copper having higher affinities for the sorbent material. Ideally, a further reduction in metal-binding capacity of the hybrid material may help for reduction feed sample volume. Figures 6.18a and 6.19 show that following withdrawal of zinc from the feed, pH goes back to the alkaline zone which confirms the sorbent's ability to respond to fluctuations of toxic metal concentrations in the influent. In addition, the sorbent bed previously loaded with the toxic metal may be useful toward reducing detection time or sample volume requirements.

6.6.1   ALTERNATE CYCLE OF SAMPLE WITH AND WITHOUT TOXIC METAL (Zn)

The exit pH profile for the column run through HIM-X for alternate cycle of feed with and without toxic heavy metal (e.g., zinc) after initial breakthrough is displayed in Figure 6.32. After the initial pH drop, the influent was replaced by feed containing only background electrolytes but no zinc (or other heavy metal) and the pH at the column exit quickly rose to 8.7. At this point again with introduction of feed sample containing trace zinc with other background electrolyte, the exit pH dropped within some 100 BVs. For a feed with 0.5 mg/L of zinc starting with a fresh HIM-X bed near about 5000 BVs of sample solution is passed to observe the first pH drop as opposed to a HIM-X bed already preloaded with zinc, which needs about 400 bed volumes to show the pH change. In terms of sample volume, 400 BVs correspond to nearly 60 mL of samples for the experimental conditions, that is, SLV, EBCT, and soon, applicable for the present investigations.

6.6.2   DETECTION OF TRACE LEAD USING HIM-X PRELOADED WITH ZINC

In order to achieve a faster detection of trace lead, HIM-X preloaded with zinc was used. Influent containing 1 mg/L zinc including common electrolytes (Na^+, Ca^{2+}, Cl^-, etc.) was passed through a column containing a tiny amount (less than 200 mg) of HIM-X. As the pH dropped, zinc laden influent was replaced by feed with common electrolytes only (no zinc) and the pH rose to more than 8.6. Subsequently, feed with trace lead (55 µg/L) was introduced and a significant drop in pH was obtained within about 350 BV, which corresponds to approximately 50 mL of solution. Figure 6.33a exhibits the pH profile for different trace lead concentrations (55, 90, and 130 µg/L) using an HIM-X bed preloaded with zinc. In each case, the pH dropped to less than 7 with the passage of 50 mL of feed sample. Figure 6.33b shows a plot of negative derivative of the pH curve ($-dpH/dBV$) versus BV for different trace lead feeds. It is

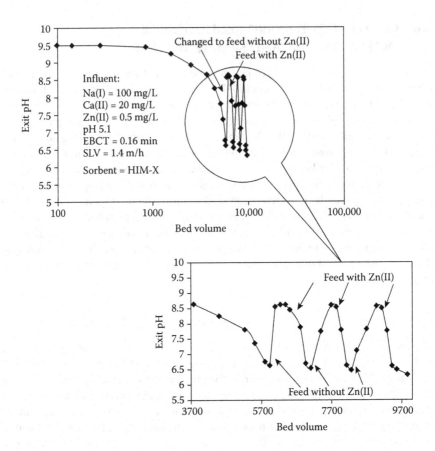

FIGURE 6.32 Exit pH profile from a fixed-bed run with HIM-X for alternate cycles with and without zinc after the initial pH drop for feed with zinc (0.5 mg/L).

worth noting that the peak heights are significantly greater than those observed for the test of Pb(II) using fresh HIM-X (Figure 6.23b), peak heights increase with an increase in lead concentration and the quantity of sample feed (hence the required time) needed to show a drop in pH reduces significantly.[50]

6.6.3 APPLICATION OF PRELOADED HIM-X IN A SYRINGE

In order to make a simple-to-operate small device that would quickly sense the presence of toxic metal(s), a syringe suitably modified for the experiment was applied. Figure 6.34a shows a syringe with a detachable tube in which a tiny amount of HIM-X (about 40 mg) is packed between two glass wool plugs. The sorbent (HIM-X) is preloaded with zinc by solution containing zinc concentration 1 mg/L before it is tested for detection of 0.1 mg/L lead. Figure 6.34b demonstrates a significant drop in pH (using indicator solution) in the presence of 0.1 mg/L of lead. The solution pH was checked using a pH meter as well. Note that in the absence of any toxic metal the pH was 8.7 but the pH dropped to 6.9 after passing 30 mL of 0.1 mg/L lead solution

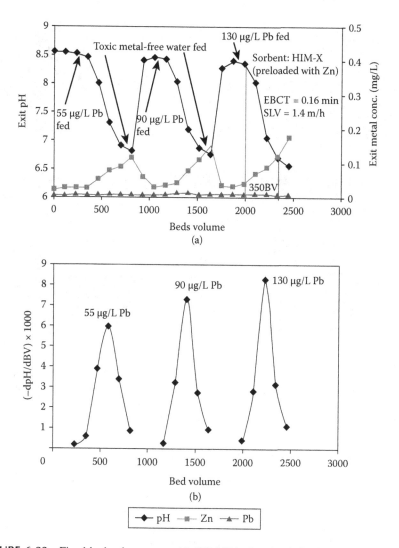

FIGURE 6.33 Fixed-bed column run with HIM-X bed preloaded with zinc. (a) Exit pH profile and zinc breakthrough with introduction of trace lead in the feed. (b) Corresponding peak heights from (−dpH/dBV) versus BV plot because of different trace lead concentrations in the feed sample.

through the modified syringe containing preloaded HIM-X.[50] Figure 6.34c is a picture of the indicator solution and reference color comparator.

The preloaded bed with limited availability of metal binding sites along with the ability of $Ca_2MgSi_2O_7$ to maintain a steady alkaline pH provide a reduction of sample volume by nearly an order of magnitude. This essentially helps detection of toxic metals through the use of a small quantity of HIM-X preloaded with Zn in a syringe explained earlier.

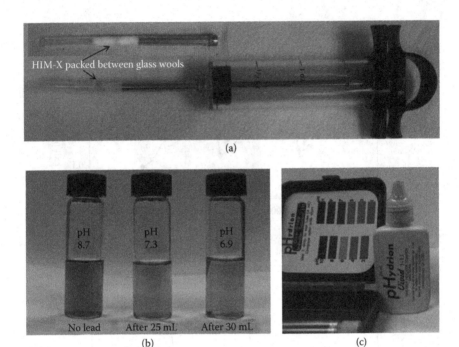

FIGURE 6.34 Test for trace lead using preloaded HIM-X in a syringe. (a) HIM-X bed sandwiched between glass wool in the detachable tube of the syringe and preloaded with zinc. (b) Demonstration of exit pH (using color comparator) from the syringe for feed without and with toxic metals (lead, 0.1 mg/L). (c) pH indicator and color comparator.

6.7 PROCESS KINETICS AND SENSING MECHANISM

The hybrid inorganic material (HIM or HIM-X) used for the detection of toxic metal through a change in pH offers simultaneous selective metal sorption and maintains a near-constant alkaline pH through dissolution or hydrolysis. The kinetics of dissolution of many sparingly soluble minerals such as calcite, silica, feldspars, and aragonite has been recognized as surface reaction controlled in the open literature.[25,37,53,73] Metal binding onto the sorbent surface is an ion exchange phenomenon that is controlled by diffusion.[25,27,54] In order to gain insight into the governing step, an "interruption test" was performed.

6.7.1 INTERRUPTION TEST—DISSOLUTION VERSUS DIFFUSION

For a column run with influent containing 0.5 mg/L zinc through HIM-X following the pH drop and zinc breakthrough (more than 25% of feed zinc concentration in the effluent) after 5400 BV, the run was deliberately stopped for 1 h. Figure 6.35a explains the result of the 1 h interruption test. When the flow was subsequently resumed, the effluent pH rose quickly from 7.5 (before interruption) to 9.3. However, after the restart and passage of about 350 BVs, the pH came down to the value prior to the interruption. The zinc concentration at the column exit dropped slightly to

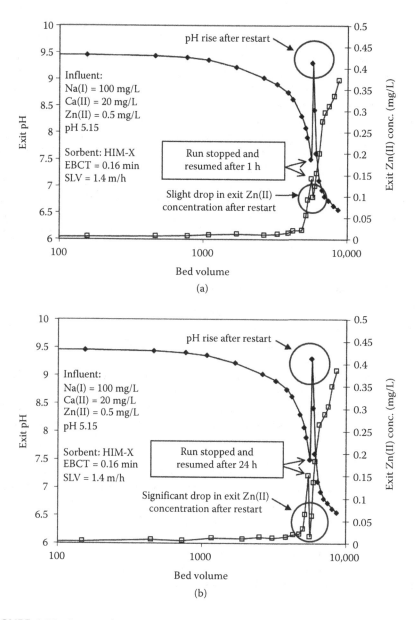

FIGURE 6.35 Interruption test after breakthrough showing its effect on dissolution and metal sorption kinetics: (a) 1 h interruption test and (b) 24 h interruption test.

0.098 mg/L after restart from 0.14 mg/L before interruption. Subsequently, a second interruption test was performed for 24 h under identical conditions after pH drop and zinc breakthrough. The result of this test is shown in Figure 6.35b. Note that as flow resumed the pH rose to 9.44 from 7.5 before the interruption and the zinc concentration at the column exit dropped to an almost negligible value compared to

28% breakthrough value of feed concentration before the interruption. As flow continued after resumption, both the pH and zinc concentration quickly reached their respective values before the stoppage of flow. The 24 h pause offered ample time for the concentration gradient to level out within sorbent particles, resulting in faster uptake and negligible zinc concentration at the exit upon restarting of the column. This observation clearly suggests the process is likely to be intraparticle diffusion controlled and it is in agreement with the results of sorption kinetics of other inorganic and organic sorbents reported in literature.[65,70,74] Comparison between the two interruption tests reveal that pH attained the equilibrium value (~9.4) following dissolution even with the 1 h interruption test, whereas a significant drop in effluent zinc concentration could be observed only with the longer time of pause, that is, 24 h interruption test. The independent dissolution test of HIM and HIM-X described in an earlier section (Figure 6.16a and b) also confirmed quick attainment of equilibrium pH. Thus, among dissolution and particle diffusion, diffusion is the slowest and is the governing rate-limiting step.

6.7.2 Intraparticle Diffusion Coefficient

A batch kinetic study was performed using a centrifugal stirrer designed for kinetic studies as demonstrated in Figure 6.36. This apparatus was originally developed by Kressman for the measurement of ion exchange rate.[55] The sorbent material (HIM or HIM-X) was placed in a cage in the central part of the stirrer. As the stirrer was immersed in the solution and started, centrifugal action produced a rapid circulating flow of solution entering through the bottom of the cage and forced out through the sorbent and radial openings provided in the stirrer. The concentration of the target solute was determined by withdrawing aliquots from the bulk solution at different times. Based upon the observation from the previous section, the kinetic study was performed under the assumption that intraparticle diffusion is the rate-limiting step. A 2 L solution containing 0.25 mg/L zinc along with the usual background of Na and Ca was placed in the container; 0.1 g of sorbent (HIM/HIM-X) was taken in the stirrer cage and the stirrer speed was kept constant at 1800 rpm. Note that the starting concentration of zinc was kept well below the solubility limit to avoid the possibility of precipitation, and the pH of the solution was maintained at around 8.8–9.0 to simulate the prevailing sorption environment.

With the assumption of spherical particles where diffusion is radial, the diffusion equation for a constant diffusivity D takes the form[75]

$$\frac{\partial Q}{\partial t} = D\left(\frac{\partial^2 Q}{\partial r^2} + \frac{2}{r}\frac{\partial Q}{\partial r}\right) \tag{6.10}$$

with r being the radial space coordinate (related to the radius of the particle), and Q is the concentration of the solute at the sorbent phase at any time t.

For a sorbent particle (spherical) that is initially free from solute, with volume of solution V, concentration of solute in the solution C_t at any time t, initial concentration C_0, the total amount of solute M_t in the sorbent at time t is expressed as a fraction

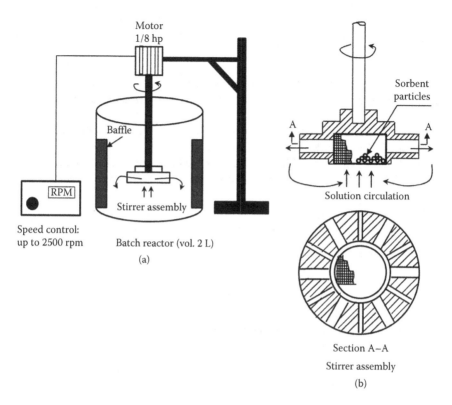

FIGURE 6.36 (a) Schematic representation of experimental set-up for batch kinetic test. (b) Schematic drawing of stirrer assembly (not to scale).

of solute uptake (F) of the corresponding quantity after infinite time (M_∞) by the following relation[70,75]:

$$F = \frac{M_t}{M_\infty} = 1 - \sum_{n=1}^{\infty} \frac{6\omega(\omega+1)\exp\left(-\frac{D\beta^2 t}{a^2}\right)}{9+9\omega+\omega^2\beta^2} \tag{6.11}$$

with βs being the non-zero roots of

$$\tan\beta = \frac{3\beta}{3+\omega\beta^2}$$

and $\omega = \dfrac{3V}{4\pi a^3}$ the ratio of the volume of the solution (V) and spherical particle with radius a. The parameter ω is expressed in terms of the final fractional uptake of solute by the spherical sorbent particle according to

$$\frac{M_\infty}{VC_0} = \frac{1}{1+\omega}$$

Mass balance gives

$$mM_t = V(C_0 - C_t) \qquad (6.12)$$

with m being the mass of sorbent used for the test, C_t the concentration of solute in the aqueous phase at time t, and M_∞ was determined from the aqueous phase concentration after equilibration for 72 h.

Figure 6.37a and b show fractional zinc uptake versus time plots for the batch kinetic tests using HIM and HIM-X, respectively. Pertinent experimental conditions are also included in the same figures. The dotted lines in the figures represent the model prediction of kinetic test results, and the best-fit intraparticle diffusivity D for zinc sorption

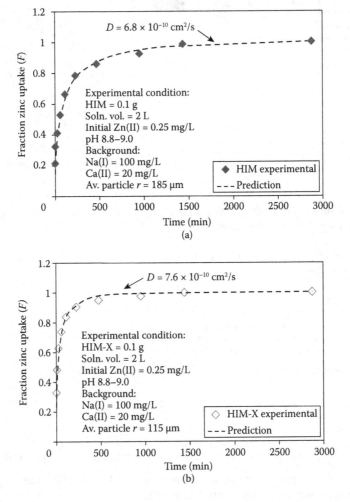

FIGURE 6.37 Results of batch kinetic test data and predicted best-fit line (dashed) with effective diffusivity. (a) Fractional zinc uptake and time plot for HIM. (b) Fractional zinc uptake and time plot for HIM-X.

onto HIM and HIM-X were computed to be 6.8×10^{-10} and 7.6×10^{-10} cm²/s, respectively. However, due to a relatively smaller particle size, HIM-X exhibits slightly higher diffusivity than HIM. The magnitude of diffusivity is consistent with the results for metal sorption by similar types of sorbents available in published literature.[46,76–78]

6.7.3 METAL SORPTION VERSUS PRECIPITATION

The experimental results displayed in Figures 6.18a, 6.19, 6.21a, and 6.23a confirm the removal of toxic metal(s) upon passage of toxic metal laden water through sorbent material (HIM or HIM-X) before occurrence of the pH drop and metal breakthrough. Conceptually, metal removal can occur by sorption onto the sorbent (through Lewis acid–base type interaction) or precipitation through the formation of metal hydroxides prompted by a near-constant alkaline pH condition (~9.0) of the sorbent surroundings. However, removal through precipitation is possible for metal concentrations above the solubility limit at the operating pH condition. Many of these metals are toxic at concentrations far below their solubility limit. The present study focuses on the influent toxic metal concentrations well below their solubility limit for lead and zinc at pH ~9.0, calculated from the data available in the open literature[11] and indicated in Table 6.6. Thus, metal removal for the present investigation can be attributed to the sorption process only. Since the activation energy for the precipitation reaction is generally much greater than the sorption process (Lewis acid–base interaction), the metal removal through sorption is kinetically preferred.[37] Experimental results show (Figures 6.18a, 6.19, 6.21a, and 6.23a) almost negligible (close to zero) metal concentration at the column exit before breakthrough. Such complete metal removal, which is way below their solubility limits, clearly suggests that sorption is the predominant removal mechanism for the two sorbents (HIM and HIM-X). This observation is in agreement with the previous investigations on similar types of sorbents.[45,46]

The sorbent materials HIM or HIM-X are sparingly soluble in water and they undergo slow hydrolysis (through chemisorptions of water molecules) in contact

TABLE 6.6
Solubility of Different Heavy Metals (Hydroxides) of Interest

Species	K_{sp} [M(OH)₂(s)]	Total Solubility, $[M]_T = [M^{2+}] + \sum M(OH)_n^{2-n}$		Influent Metal Concentration in Experiments
		pH 5	pH 9.0	
Pb^{2+}	$10^{-15.3}$	1.04×10^5 g/L	30 mg/L	≤0.5 mg/L
Zn^{2+}	$10^{-15.5}$	2.06×10^4 g/L	3.02 mg/L	≤1 mg/L
Cu^{2+}	$10^{-19.3}$	3.19 g/L	3.04×10^{-3} mg/L	≤0.5 mg/L

Source: Stum, W., and Morgan, J. J.: *Aquatic Chemistry: Chemical Equilibria and Rates in Natural Waters,* Third Edition. 1995. Copyright Wiley-VCH Verlag GmbH & Co. KGaA. Reproduced with permission.

Note: K_{sp} and M refer solubility product and representative metal.

with water. The surface functionalities of oxides and /or silicates present in HIM and HIM-X are modified and covered with surface hydroxyl groups that can act as both proton donors and acceptors (refer to Figure 6.3). Several previous studies on the sorption of heavy-metal cations and ligands on different metal-oxides/silicate surfaces have validated this premise.[53] Blum and Lasaga[79] have attributed surface binding on aluminum sites (Al(OH)) for albite (a mineral with primary constituent $AaAlSi_3O_8$). In sepiolite (primary component $Mg_4Si_5O_{15}(OH)_2 \cdot 6H_2O$) Vico[28] categorized $\equiv MgOH$ and $\equiv AlOH$ as the predominant heavy-metal binding sites. Similar validations have been reported through previous investigations available in the published literature.[80–84]

The surface hydroxyls formed for different oxides' surfaces may not be fully structurally and chemically equivalent, but for a schematic representation of the reactions, it may be written as: $\equiv S–OH$. Selective metal (e.g., Zn^{2+}) binding may be represented by the reaction

$$2\equiv S–OH + Zn^{2+} \leftrightarrow (S–O)_2Zn + 2H^+ \tag{6.13}$$

Such metal binding is pH dependent and greatly enhanced at pH > pHzpc.

6.7.4   EDX ANALYSES OF SORBENTS HIM AND HIM-X

EDX studies reveal the chemical characterization showing qualitative elemental analysis of the surface of samples.[85] Figure 6.38a and b show the EDX analysis for a fresh and a used HIM sample after passing a solution containing trace zinc (0.5 mg/L or 7.6×10^{-6} M) and other common electrolytes (Na^+ = 100 mg/L or 1.7×10^{-3} M, Ca^{2+} = 20 mg/L or 5×10^{-4} M), respectively. Peaks of different major elements present at the surface of the fresh HIM particle such as O, Fe, Si, Ca, and Mg (Figure 6.38a) are also available from used HIM particles (Figure 6.38b) except for zinc, which is significantly present in the used HIM but absent in the fresh HIM. Reduction in peaks of Si, O, Ca, and Mg are found in used HIM as a result of dissolution, whereas the peak for Fe remained almost unaltered in both the samples. In a similar approach, EDX analysis of a fresh and a used HIM-X sample is shown in Figure 6.39a and b. A comparison of peaks reveals zinc is only available in used HIM-X, and peaks of O, Si, Ca, and Mg though found in both the samples are especially reduced in the used sample as a result of loss due to dissolution during the experimental run. Note that little sodium is present in either of the used HIM and HIM-X. However, comparing influent Zn concentration that is less by more than 2 orders of magnitude to the Na concentration, Zn uptake on HIM/HIM-X surfaces is quite significant. This observation corroborates the premise of selective binding of toxic metal cations (zinc, i.e., Zn^{2+} in this case) on the surface of HIM and HIM-X and sorption is the predominant mechanism for removal. Such selective toxic metal sorptions by different metal oxide surfaces are well supported by published literature.[25,27,29,30,42,58,78,81] However, Na and Ca, which were present at much higher concentration than Zn in the influent, might bind onto the HIM/ HIM-X surface through weak electrostatic interactions causing a slight addition to their peaks in the used sorbents.

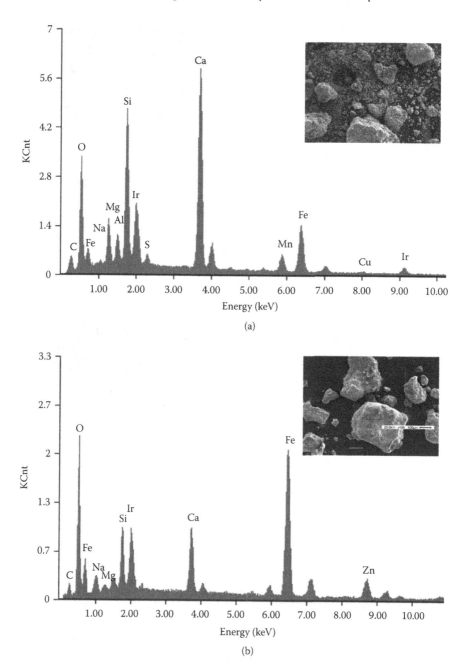

FIGURE 6.38 EDX spectra showing elemental analysis of (a) fresh HIM sample and (b) used HIM sample retrieved after column experiment using feed containing trace zinc (0.5 mg/L).

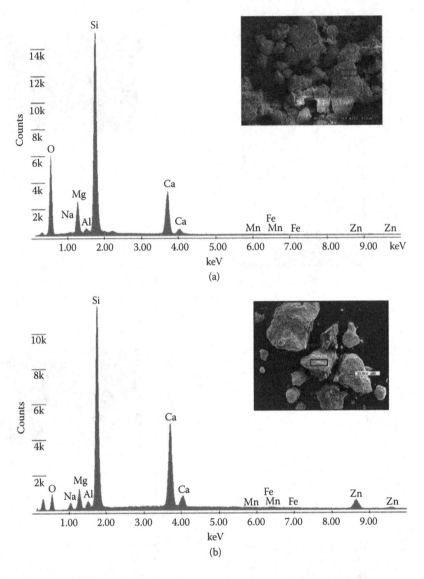

FIGURE 6.39 EDX spectra showing elemental analysis of (a) fresh HIM-X sample and (b) used HIM-X sample retrieved after column experiment using feed containing trace zinc (0.5 mg/L).

6.7.5 Toxic Metal Sensing Mechanism

The experimental results demonstrated that toxic metals, when present in water at concentrations well below 1.0 mg/L, can be detected through pH changes occurring at the exit during an HIM mini-column run. More significantly, the slope of the exit pH curve (i.e., −dpH/dBV) provides distinguishable peaks, confirming the presence of toxic metals, namely, zinc and lead. The hybrid inorganic sorbent materials HIM and HIM-X used for the present studies offer unique sorption for toxic metals while maintaining a near-constant slightly alkaline pH for a prolonged period. Akermanite in HIM or a calcium magnesium silicon oxide composite in HIM-X in contact with water slowly dissolutes, releasing hydroxyl ions that essentially maintain a steady pH around 9.0 according to the following reaction (the over bar indicates solid phase):

$$\overline{Ca_2MgSi_2O_7} + 3H_2O \leftrightarrow 2Ca^{2+} + Mg^{2+} + 2SiO_2 + 6OH^-$$

The akermanite or calcium magnesium silicate is responsible for maintaining a near-constant alkaline pH through hydrolysis for a very prolonged period of time in accordance with the above reaction. Batch dissolution studies with HIM and HIM-X also verify that it can maintain almost constant pH for a prolonged period through slow hydrolysis (Figure 6.16a and b). Besides, the rise of pH to the alkaline domain following the withdrawal of zinc (Figures 6.18a, 6.19, 6.32, and 6.33a) from the feed confirms the ability of HIM/HIM-X to respond to fluctuations in toxic metal concentrations in the influent sample and the prolonged dissolution characteristics of $Ca_2MgSi_2O_7$. The protolytic reactions of hydroxylated surface of HIM or HIM-X commonly may be represented by

$$\equiv S{-}OH^{2+} \leftrightarrow \equiv S{-}OH + H^+ \tag{6.14}$$

$$Ka_1 = \{\equiv S{-}OH\}[H^+]/\{\equiv S{-}OH^{2+}\} \tag{6.15}$$

$$\equiv S{-}OH \leftrightarrow \equiv S{-}O^- + H^+ \tag{6.16}$$

$$Ka_2 = \{\equiv S{-}O^-\}[H^+]/\{\equiv S{-}OH\} \tag{6.17}$$

Though measurement/estimation of equilibrium constants for surface protonation/ de-protonation (Ka_1 and Ka_2 in Equations 6.15 and 6.17) of HIM was not conducted under the present study, zeta potential values at different ionic strengths shown in Table 6.7 well confirm the overall negative surface charge at equilibrium pH of the experiment (≥ 9.5). At this alkaline pH, the predominant surface functional group ($\equiv S{-}O^-$) with oxygen donor atom offers a strong binding affinity for Pb^{2+}, Cu^{2+}, Zn^{2+}, and other transition metal cations forming inner sphere complexes through Lewis acid–base interactions. In comparison, common alkaline and alkaline-earth metal cations (Na^+, Ca^{2+}) form only outer sphere complexes and have poor affinity for the $\equiv S{-}O^-$ binding sites.[29,53]

All the toxic heavy metals (e.g., Pb, Cu, Zn) form labile metal–hydroxy complexes at alkaline pH in accordance with the following reaction (M refers to the metal):

$$M^{2+}(aq) + nOH^- \leftrightarrow (M(OH)_n)^{+2-n}$$

$$K_n \text{ (Equilibrium or stability constant)} = [(M(OH)_n)^{+2-n}] / [M^{2+}] [OH^-]^n \tag{6.18}$$

For $n = 1, 2, 3$.

On the contrary, common innocuous cations such as Na^+ and Ca^{2+} form poor complexes with OH^- ions. Table 6.8 provides the first and second stability or association constants ($\log K_1$ and $\log K_2$) for different metals of interest.[86] The stability or association constant (K_1) for metal-hydroxy complex formation for these toxic metals is different and several orders of magnitude higher compared to that of common alkaline and alkaline-earth metal cations. The high values of stability constant of these metals indicate that they may cause significant pH reduction, even below micromolar concentrations, by forming metal-hydroxy complexes.

A sample containing a toxic metal such as zinc including other common electrolytes is passed through a HIM-X (or HIM) bed for sensing of zinc through pH change. However, the sample is made free of bicarbonate through adjustment of pH to ≤5.0 with N_2 purging (Figure 6.26). Subsequently, other buffering solutes such

TABLE 6.7

Zeta Potential at Different Molar Concentrations (Using NaNo₃ Solution) for Fresh HIM Particles and HIM Particles Loaded with Zinc (Measured at Equilibrium pH ~9.5)

Molar Concentration of NaNO₃ Solution (M)	Zeta Potential (mV)	
	HIM Fresh	HIM-Zn Loaded
0.0001	−40.2	−25.5
0.001	−39.5	−24.8
0.01	−35.7	−22.4
0.1	−28.8	−19.8
1.0	−18.7	−14.1

Source: Chatterjee, P. K., *Sensing and Detection of Toxic Metals in Water with Innovative Sorption Based Techniques,* PhD Thesis, Civil and Environmental Engineering, Lehigh University, Bethlehem, PA, 2010.

TABLE 6.8

Stability Constants of Metal–Hydroxy Complexes for Different Toxic Metals of Interest

Metal Ion	Stability Constants for Metal–Hydroxy Complex	
	Log K_1	Log K_2
Zn^{2+}	5.0 [$Zn(OH)^+$]	11.1 [$Zn(OH)_2$]
Pb^{2+}	6.3 [$Pb(OH)^+$]	11.0 [$Pb(OH)_2$]
Cu^{2+}	6.3 [$Cu(OH)^+$]	11.8 [$Cu(OH)_2$]

Source: Martell, A. E., and Smith, R. M., *Critical Stability Constants*, Vols. 1–6, Plenum Press, New York, 1977.

as phosphates, NOM, or other anionic ligands, if present, are removed by passing through HAIX media ahead of HIM-X as explained in Figure 6.30a and b. Due to the Donnan co-ion exclusion effect, Zn^{2+} and other cations are completely rejected by the HAIX (Figure 6.29). On the contrary, an anionic ligand (e.g., phosphate or fulvate in NOM) permeates easily into the exchanger phase of the HAIX and gets selectively bound onto the sorption sites of HFO in preference to other commonly occurring anions, namely, sulfate and chloride as explained in Figure 6.31. As the sample percolates through the sorbent HIM-X, the pH turns alkaline (>9.0) due to hydrolysis of the silicate phase according to the reaction in Equation 6.2. If any toxic solute (say zinc) is present in the sample at trace concentration, Zn^{2+} is selectively bound to the surface binding sites while other electrolytes (Na^+, Ca^{2+}) pass through and the exit pH still remains alkaline. Finally, as the sorption sites get exhausted, Zn^{2+} exits from the HIM or HIM-X bed, instantaneously forming complexes with OH^- causing a drop in pH.

The sequence of steps leading to the sharp drop in pH in the presence of trace toxic metal ions is illustrated in Figure 6.40. For every column run when the exit pH remained alkaline, toxic metal concentrations at the exit of the column were non-detectable. The pH dropped concurrently with the breakthrough of metal ions (Figures 6.18a and b; 6.21a and b; 6.22a, b, and c; 6.23a and b; and 6.24a and b). This observation suggests that, upon saturation of binding sites, metals (e.g., zinc) break through and rapidly form metal hydroxyl complexes, resulting in the depletion of hydroxyl ions and consequent drop in pH. Once the toxic metals are withdrawn from the feed, the pH again swings back to alkaline domain (Figures 6.19, 6.32, and 6.33a). In principle, the Donnan co-ion exclusion effect of HAIX, hydrolysis of calcium magnesium silicate leading to slow release of OH^- for a very prolonged period, and the specific metal binding sites ($\equiv S{-}O^-$) together create a synergy that allows pH to act as a surrogate indicator to sense toxic metals overcoming interferences of phosphate/NOM. A mathematical correlation between aqueous phase metal ion and hydroxyl ion concentration may be expressed as

$$[M]_T = [M^{2+}] + \sum [M(OH)_n^{+2-n}] \tag{6.19}$$

$$[OH^-]_T = [OH^-] + \sum [M(OH)_n^{+2-n}] \tag{6.20}$$

where $[M]_T$ and $[OH^-]_T$ are total metal and OH^- ion concentrations, $[M^{2+}]$ is free metal ion concentration (not in complex), and $[M(OH)_n^{2-n}]$ are concentrations of different metal–hydroxy species.

Using appropriate stability constants, the drop in pH can be calculated for different metal breakthrough concentrations. Previous studies by Gao[46] and Kney and SenGupta[45] at Lehigh University have validated the scientific premise of a drop in aqueous phase pH due to the formation of metal–hydroxy complexes through experimental observations complemented by theoretical calculations.

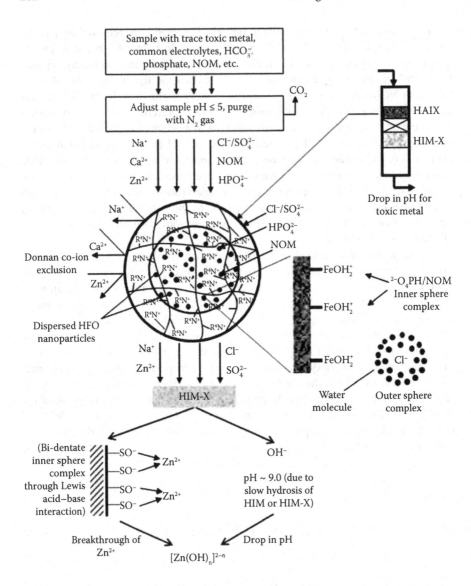

FIGURE 6.40 Schematic illustration of toxic metal–sensing mechanism through change of pH with a modified bed of HAIX and HIM-X for avoiding interference of buffers, such as bicarbonates, phosphate, and NOM.

6.7.6 METAL HYDROXY COMPLEX FORMATION AND SPECIATION DISTRIBUTION

The toxic metal cations such as Pb(II), Cu(II), and Zn(II) included in this study belong to either the soft or the borderline soft category according to HSAB rules. They exhibit strong coordination interaction with a variety of inorganic and organic ligands and form metal–ligand complexes.[4–6,11,12] The extent of complex formation primarily depends on the concentration of metals, concentration of ligands, metal ligand affinity, strength of complex (i.e., formation or stability constant), system pH, and ionic strength. These toxic metal cations form strong complexes with hydroxyl ions as evidenced from Table 6.8. Figure 6.41 demonstrates the speciation distribution of metal and metal hydroxy complexes as a function of pH for Pb. The speciation distribution diagrams are developed based on the stability constants available in literature[37,86] according to the calculations for Pb^{2+} elaborated below.

The different lead–hydroxy species considered for calculation of speciation distributions based on Equation 6.18 are

$$[Pb(OH)^+] = K_1 [Pb^{2+}] [OH^-] \tag{6.21}$$

$$[Pb(OH)_2] = K_2 [Pb^{2+}] [OH^-]^2 \tag{6.22}$$

$$[Pb(OH)_3^-] = K_3 [Pb^{2+}] [OH^-]^3 \tag{6.23}$$

Following Equation 6.19, the total concentration of dissolved Pb^{2+} is given by

$$[Pb^{2+}]_T = [Pb^{2+}] + [Pb(OH)^+] + [Pb(OH)_2] + [Pb(OH)_3^-] \tag{6.24}$$

Replacing Equations 6.21, 6.22, and 6.23 in Equation 6.24,

$$[Pb]_T = [Pb^{2+}] + K_1 [Pb^{2+}] [OH^-] + K_2 [Pb^{2+}] [OH^-]^2 + K_3 [Pb^{2+}] [OH^-]^3 \tag{6.25}$$

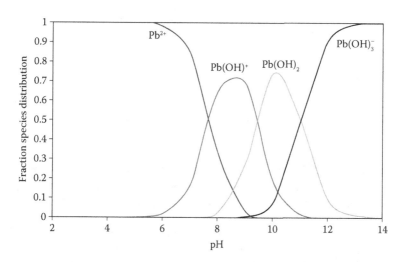

FIGURE 6.41 Speciation distributions of free metal cation and metal hydroxy complexes as a function of pH for lead and lead-hydroxy species (calculated based on the data available from the open literature). (From Stum, W., and Morgan, J. J: *Aquatic Chemistry: Chemical Equilibria and Rates in Natural Waters,* Third Edition. 1995. Copyright Wiley-VCH Verlag GmbH & Co. KGaA. Reproduced with permission; Sposito, G., *The Surface Chemistry of Soils,* 1984, by permission of Oxford University Press; Martell, A. E., and Smith, R. M., *Critical Stability Constants,* Vols. 1–6, Plenum Press, New York, 1977.)

Values of Ks, that is, stability constants available in literature,[86] are

$$K_1 = 10^{6.3}, K_2 = 10^{11}, K_3 = 10^{14}$$

Therefore, rearranging the above equation gives the fraction of free Pb^{2+} species:

$$\frac{\left[Pb^{2+}\right]}{\left[Pb^{2+}\right]_T} = \frac{1}{1 + K_1[OH^-] + K_2[OH^-]^2 + K_3[OH^-]^3} \tag{6.26}$$

In a similar way, expressing $[Pb^{2+}]$ and all $[Pb(OH)_n^{2-n}]$ species in Equation 6.25 in the form of $[Pb(OH)^+]$ and using relations of Equations 6.21, 6.22, and 6.23 in Equation 6.25 gives the fraction of $[Pb(OH)^+]$ species:

$$\frac{\left[Pb(OH)^+\right]}{[Pb]_T} = \frac{1}{\dfrac{1}{K_1[OH^-]} + 1 + \dfrac{K_2}{K_1}[OH^-] + \dfrac{K_3}{K_1}[OH^-]^2} \tag{6.27}$$

Similarly, the fraction of species $[Pb(OH)_2]$ and $[Pb(OH)_3^-)]$ may be expressed through the following equations:

$$\frac{\left[Pb(OH)_2\right]}{[Pb]_T} = \frac{1}{\dfrac{1}{K_2[OH^-]^2} + \dfrac{K_1}{K_2[OH^-]} + 1 + \dfrac{K_3}{K_2}[OH^-]} \tag{6.28}$$

$$\frac{\left[Pb(OH)_3^-\right]}{[Pb]_T} = \frac{1}{\dfrac{1}{K_3[OH^-]^3} + \dfrac{K_1}{K_3[OH^-]^2} + \dfrac{K_2}{K_3[OH^-]} + 1} \tag{6.29}$$

The fraction contribution of different species at any pH can be estimated using Equations 6.26, 6.27, 6.28, and 6.29. The fraction of free Pb^{2+} and $Pb(OH)_n^{2-n}$ species at two different pHs, namely pH 9.0 and pH 7.0, calculated based on the above equations are given Table 6.9.[50]

It explains that at slightly alkaline pH (~9), the predominant species are $Pb(OH)^+$ and $Pb(OH)_2$. The fraction of free Pb^{2+} is negligible compared to lead–hydroxy complexes. The speciation distribution diagram of lead and lead–hydroxy complexes as a function of pH displayed in Figure 6.41 also confirms that the formation of metal hydroxy complexes at slightly alkaline pH would result in a drop in pH. In a similar vein, speciation distribution diagrams for copper and zinc hydroxy species as a function of pH would confirm the predominance of metal–hydroxy complexes at pH around 9 that essentially causes a drop in pH in the aqueous phase through sequestration of OH^- ions.

TABLE 6.9

Speciation Distribution of Free Lead Ion (Pb^{2+}) and Lead–Hydroxy Complexes at pH 7 and 9

pH	Fraction [Pb^{2+}]	Fraction [Pb(OH)$^+$]	Fraction [Pb(OH)$_2$]	Fraction [Pb(OH)$_3^-$]
9.0	0.032	0.693	0.272	0.003
7.0	0.834	0.165	0.001	—

6.8 CONCLUSIONS

This present study validates the concept of sensing of toxic metals dissolved in water using pH as the sole surrogate parameter through application of hybrid inorganic sorbent materials (HIM or HIM-X), which are basically a granular composite of calcium–magnesium–silicate (Ca$_2$MgSi$_2$O$_7$) with or without a small amount of HFO. From the water quality viewpoint, the present technique offers simple diagnostic tests for different toxic metals such as lead, zinc, and copper in water only through pH change. The process is not interfered by other common electrolytes, which are usually present at much higher concentrations. For typical surface water and municipal water systems, the technique seems to be quite effective for detecting toxic metals at concentrations well below 100 µg/L. The major findings of this study may be summarized as follows:

- The exit pH from a fixed-bed column run maintains steady alkaline pH (~9.0) for influent containing common electrolytes (Na$^+$, Ca^{2+}, Cl$^-$, SO$_4^{2-}$). In contrast, feeds containing trace amounts of Zn(II)/Pb(II)/Ni(II)/Cu(II) in addition to common electrolytes show significant pH drops (>2 units) at the column exit after a lag time. Upon withdrawal of toxic metal cations from the feed, pH at the column exit switches back to the alkaline domain.
- In the presence of toxic metals, the pH drop coincides with metal breakthrough. A plot of the negative derivative of effluent pH profile (−dpH/dBV) versus bed volume (BV) produces a peak that is characteristic of the type and concentration of toxic metal in the feed solution.
- Akermanite or calcium magnesium silicate, which is sparingly soluble in water, slowly hydrolyzes releasing OH$^-$ ions through incongruent dissolution and maintains a near-constant alkaline pH (~9.0) for a prolonged period. At this pH condition, surface hydroxyl ($\equiv$S–OH) of HIM/HIM-X upon de-protonation selectively binds toxic metal cations. Upon saturation of binding sites, toxic metal cations break through and instantaneously form metal–hydroxy complexes resulting in a pH change caused by the dissipation of OH$^-$ ions from the aqueous phase.
- Adjustment of the inlet pH to <5.0 and subsequent N$_2$ purging allows stripping off bicarbonate (and carbonate as well) in the form of CO$_2$ from the aqueous phase and overcomes interference. Use of HAIX prior to HIM overcomes interference of phosphate and NOM (commonly encountered

buffering species) in different types of water. In HAIX, HFO nanoparticles dispersed within the polymeric phase selectively remove phosphate or other weak-acid anionic ligands (e.g., fulvate in NOM) prompted by the Donnan membrane effect due to the presence of non-diffusible quaternary ammonium (R_4N^+) functional groups in the polymeric phase. This technique validates the detection of lead and zinc in synthetic water, tap water, and river water.

- The sensitivity of the technique for the detection of ultra-low concentrations of toxic metals (e.g., lead) or other metals (less than 10 µg/L) is validated using an operationally simple preconcentration method.

- Use of HIM-X presaturated with zinc enables sensing of toxic metals (e.g., lead) with a significant reduction in sample volume. The preloaded bed with limited available metal-binding sites along with the ability of $Ca_2MgSi_2O_7$ to maintain a steady alkaline pH provides a reduction of sample volume by nearly an order of magnitude.

- Toxic metal concentrations in the feed at levels well below their solubility limits essentially proves that sorption is a predominant metal (cations) removal mechanism before breakthrough and the observed pH drop. Kinetic studies suggest that intraparticle diffusion is the governing rate controlling step.

- Since pH is the sole surrogate indicator, except for a pH meter and/or indicator solution, no sophisticated instrument or chemicals are required. The unique sorption and dissolution behavior of HIM and HIM-X and the ability of toxic metals to form labile metal hydroxy complexes at slightly alkaline pH offer a synergy to cause a sharp pH drop in the presence of toxic metals even below micromolar concentration. A sharp drop in pH (around 2 units or more) signals the presence of one or more toxic metals, which may be a useful primary identification tool for toxic metal contamination in water. According to information in the open literature, no current techniques use pH as the sole parameter to sense the presence of toxic metals in water.

REFERENCES

1. SenGupta, A. K., Ed. *Environmental Separation of Heavy Metals—Engineering Processes*. Lewis, New York, 2002.
2. Davis, M. L., Cornwell, D. A. *Introduction to Environmental Engineering*, Third Edition. McGraw Hill, New York, 1998.
3. U.S. E.P.A, *National primary drinking water regulations–list of drinking water contaminants and their MCLs*. Available from http://www.epa.gov/safewater/mcl.html
4. Duffus, J. H. "Heavy metals"—A meaningless term? *Pure Appl. Chem.* **2002**, *74* (5), 793–807.
5. Pearson, R. G. Hard and soft acid and bases, HSAB, Part 1. *J. Chem. Ed.* **1968**, *45* (9), 581–587.
6. Pearson, R. G. Hard and soft acids and bases, HSAB, Part 2. *J. Chem. Ed.* **1969**, *45* (10), 643–648.
7. Flemming, C. A., Trevors, J. T. Copper toxicity and chemistry in the environment— A review. *Water Air Soil Pollut.* **1989**, *44*, 143–158.
8. Hobbs, P. V., Harison, H., Robinson, E. The biological cycle of mercury. *Science* **1974**, *183*, 909–915.

9. Hayes, A. W. *Principles and Methods of Toxicology.* CRC Press, Philadelphia, 2007.

10. Lane, T. W., Saito, M. A., George, G. N., Pickering, I. J., Prince, R. C., Morel, F. M. M. A cadmium or enzyme for marine diatom. *Nature* **2005**, *435*, 42.

11. Stum, W., Morgan, J. J. *Aquatic Chemistry: Chemical Equilibria and Rates in Natural Waters,* Third Edition. Wiley-Interscience, New York, 1995.

12. Nieober, E., Richardson, D. H. S. The replacement of the non-descript term "Heavy Metals" by a biologically and chemically significant classification metal ions. *Environ. Pollut. Ser. B.* **1980**, *1*, 3.

13. Wang, J. Remote electro-chemical sensors for monitoring inorganic and organic pollutants. *Trends Anal. Chem.* **1997**, *16* (2), 84–88.

14. Schöning, M. J., Hullenkremer, B., Glück, O., Lüth, H., Emons, H. Voltohmmetry— A novel sensing principle for heavy metal determination in aqueous solutions. *Sensor. Actuator. B. Chem.* **2001**, *76* (1–3), 275–280.

15. Ju M.-J., Hayashi, K., Toko, K., Yang, D. H., Lee, S. W., Kunitake, T. A new electro-chemical sensor for heavy metal ions by the surface polarization controlling method. In *The 13th International Conference on Solid-State Sensors Actuators and Microsystems,* Seoul, Korea, 2005.

16. Eicken, C., Pennella, M. A., Chen, X., Koshlap, K. M., VanZile, M. L., Sacchettini, J. C., Giedroc, D. P. A metal-ligand mediated intersubunit allosteric switch in related SmtB/ArsR zinc sensor proteins. *J. Mol. Biol.* **2003**, *333*, 683–695.

17. Xiao, Y., Rowe, A. A., Plaxco, K. W. Electrochemical detection of parts per billion lead via an electrode-bound DNAzyme assembly. *J. Am. Chem. Soc.* **2007**, *129*, 262–263.

18. Bambang, K. Optical chemical sensors for the determination of heavy metal ions: A mini review. *J. ILMU DASAR* **2000**, *1* (2), 18–29.

19. Gao, H. W., Chen, F., Chen, L., Zeng, T., Pan, L., Li, J. H., Luo, H. F. A novel detection approach based on chromophore-decolorizing with free radical and application to photometric determination of copper with acid chrome dark blue. *Anal. Chim. Acta.* **2007**, *587* (1), 52–59.

20. Sumner, J. P., Westerberg, N. M., Stoddard, A. K., Hurst, T. K., Cramer, M., Thompson, R. B., Fierke, C. A., Kopelman, R. DsRed as a highly sensitive, selective, and reversible fluorescence-based biosensor for both Cu+ and Cu2+ ions. *Biosens. Bioelectron.* **2006**, *21* (7), 1302–1308.

21. Forzani, E. S., Zhang, H. Q., Chen, W., Tao, N. J. Detection of heavy metal ions in drinking water using a high resolution surface plasmon resonance sensor. *Environ. Sci. Technol.* **2005**, *39* (5), 1257–1262.

22. Arduini, M., Mancin, F., Tecilla, P., Tonellato, U. Self-organized fluorescent nanosensors for ratiometric Pb2+ detection. *Langmuir.* **2007**, *23*, 8632–8636.

23. Parks, G. A. Surface energy and adsorption at mineral/water interface: An introduction. In *Mineral-Water Interface Geochemistry,* Hochella, M. F., Jr., White, A. F., Eds. Mineralogical Society of America, Washington, DC, 1990, pp. 133–175.

24. Wolfgang, H. H. *Ion Exchange: Exchangers, Fundamentals, Applications, Technology.* Forschungszentrum Karlsruhe Institute of Technical Chemistry, Germany, 2008.

25. Brown, G. E., Heinrich, V. E., Casey, W. H., Clark, D. L., Eggleston, C., Felmy, A., Goodman, D. W., et al. Metal oxide surfaces and their interactions with aqueous solution and microbial organism. *Chem. Rev.* **1999**, *99* (1), 77–174.

26. Cowan, C. E., Zactora, J. M., Resch, C. T. Cadmium adsorption on iron oxides in the presence of alkaline-earth elements. *Environ. Sci. Technol.* **1991**, *25*, 437–443.

27. Tamura, H., Katayama, N., Furuichi, R. Modeling of ion-exchange reactions on metal oxides with the Frumkin isotherm. 1. Acid-base and charge characteristic of MnO_2, TiO_2, Fe_3O_4 and Al_2O_3 surfaces and adsorption affinity of alkali metal ions. *Environ. Sci. Technol.* **1996**, *30*, 1198–1204.

28. Vico, L. I. Acid-base behaviour and Cu2+ and Zn2+ complexation properties of the sepiolite/water interface. *Chem. Geol.* **2003**, *198*, 213–222.
29. Dzombak, D. A., Morel, F. M. M. *Surface Complexation Modeling: Hydrous Ferric Oxides*. Wiley-Interscience, New York, 1990.
30. Gao, Y., SenGupta, A. K., Simpson, D. A. New hybrid inorganic sorbent for heavy metals removal. *Water Res.* **1995**, *29* (9), 2195–2205.
31. Cortina, J., Lagreca, I., Pablo, J. D. Passive in-situ remediation of metal-polluted water with caustic magnesia: Evidence from column experiments. *Environ. Sci. Technol.* **2003**, *37*, 1971–1977.
32. Rotting, T. S., Cortina, J. L., Pablo, J. D. Use of caustic magnesia to remove cadmium, nickel and cobalt from water in passive treatment systems: Column experiments. *Environ. Sci. Technol.* **2006**, *40*, 6438–6443.
33. Jachova, M., Puncochar, M., Horacek, J., Stamberg, K., Vopalka, D. Removal of heavy metals from water by lignite-based sorbent. *Fuel.* **2004**, *83*, 1197–1203.
34. Rayner-Canham, G. *Descriptive Inorganic Chemistry*, Second Edition. W.H. Freeman, New York, 2000.
35. Benjamin, M. M., Leckie, J. J. Multiple sites adsorption of Cd, Cu, Zn and Pb on amorphous iron oxy hydroxides. *J. Colloid. Interface Sci.* **1981**, *79*, 209–221.
36. Sposito, G. *The Surface Chemistry of Soils*. Oxford University Press, New York, 1984.
37. Morel, F. M. M. *Principles of Aquatic Chemistry*. Wiley Interscience, New York, 1983.
38. Chatterjee, P. K., SenGupta, A. K. Sensing of toxic metals through pH changes using a hybrid sorbent material: Concept and experimental validation. *AIChE J.* **2009**, *55* (11), 2997–3004.
39. Chatterjee, P. K., SenGupta, A. K. Toxic metal sensing through pH changes using a novel hybrid sorbent material. In *ACS 240th National Symposium*, August 24, Boston, MA, 2010.
40. Sarkar, S., Chatterjee, P. K., Cumbal, L. H., SenGupta, A. K. Hybrid ion exchanger supported nanocomposites: Sorption and sensing for environmental applications. *Chem. Eng. J.* **2011**, *166*, 923–931.
41. SenGupta, A. K., Chatterjee, P. K. Rapid Sensing of Toxic Metals with Hybrid Inorganic Materials. U. S. Patent 8,187,890, May 29, 2012.
42. Du, Q., Sun, Z., Forsling, W., Tang, H. Adsorption of copper at aqueous illite surfaces. *J. Colloid Interface Sci.* **1997**, *187*, 232–242.
43. Zhuang, Y., Yang, Y., Xiang, G., Wang, X. Magnesium silicate hollow nanostructures as highly efficient adsorbents for toxic metal ions. *J. Phys. Chem. C.* **2009**, *113*, 10441–10445.
44. Green-Ruiz, C. Effect of salinity and temperature on the adsorption of Hg(II) from aqueous solutions by a Ca-montmorillonite. *Environ. Technol.* **2009**, *30* (1), 63–68.
45. Kney, A. D., SenGupta, A. K. Synthesis and characterization of a new class of a hybrid inorganic sorbents for heavy metals removal. In *Ion Exchange and Solvent Extraction: A Series of Advance*, Marcus, Y., Ed. Marcel Dekker, New York, 2001, pp. 295–352.
46. Gao, Y. M. *Sorption Enhancement of Heavy Metals onto a Modified Iron-Rich Material*, PhD Thesis, Civil and Environmental Engineering, Lehigh University, Bethlehem, PA, 1995.
47. Richard, A. B., Kenneth, W. B., Monte, C. N., John, W. A. *Hand Book of Mineralogy: Silica, Silicates*, Mineral Data Publishing, USA, 1995.
48. Swainson, I. P., Dove, M. T., Schmahl, W. S., Putnis, A. Neutron powder diffraction study of the akermanite-gehlenite solid solution series. *Phys. Chem. Mineral.* **1992**, *19*, 185–195.
49. mindat.org. *Akermanite: Akermanite mineral information and data*. Available from http://www.mindat.org/min-70.html retrieved March 30, 2009.

50. Chatterjee, P. K. *Sensing and Detection of Toxic Metals in Water with Innovative Sorption Based Techniques.* PhD Thesis, Civil and Environmental Engineering, Lehigh University, Bethlehem, PA, 2010.

51. Chatterjee, P. K., SenGupta, A. K. Toxic metal sensing through novel use of hybrid inorganic and polymeric ion. *Solvent Extraction Ion Exchange.* **2011**, *29*, 398–420.

52. APHA, AWWA., WEF., *Standard Methods for the Examination of Water and Waste Water*, Eighteenth Edition. American Public Health Association, Washington, DC, 1992.

53. Schindler, P. W., Stum, W. The surface chemistry of oxides, hydroxides and oxide minerals, in Aquatic surface chemistry. In Stum, W., Ed. Wiley-Interscience, New York: John Wiely and Sons, 1987, pp. 83–100.

54. Weber, W. J. J. *Physicochemical Processes for Water Quality Control.* Wiley-Interscience, New York, 1972.

55. Kressman, T. R. E., Kitchener, J. A. *Discussions Faraday Soc.* **1949**, 7, 90.

56. SenGupta, A. K., Lim, L. Modeling chromate ion exchange processes. *AIChE J.* **1988**, *34* (12), 2019–2029.

57. SenGupta, A. K., Zhu, Y., Hauze, D. Metal(II) ion binding onto chelating exchangers with multiple nitrogen donor atoms: Some new observations and related implications. *Environ. Sci. Technol.* **1991**, *25* (3), 481–488.

58. Bradl, H. B. Adsorption of heavy metal ions on soils and soils constituents. *J. Colloid. Interface Sci.* **2004**, *277*, 1–18.

59. Cumbal, L., Greenleaf, J., Leun, D., SenGupta, A. K. Polymer supported inorganic nanoparticles: Characterization and environmental applications. *React. Funct. Polym.* **2003**, *54*, 167–180.

60. Jyo, A., Kugara, J., Trobradovic, H., Yamabe, K., Sugo, T., Tamada, M., Kume, T. Fibrous iminodiacetic acid chelating cation exchangers with a rapid adsorption rate. *Ind. Eng. Chem. Res.* **2004**, *43*, 1599–1607.

61. Zhao, D., SenGupta, A. K. Ultimate removal of phosphate from wastewater using a new class of polymeric ion exchangers. *Water Res.* **1998**, *32* (5), 1613–1625.

62. Blaney, L. M., Cinar, S., SenGupta, A. K. Hybrid anion exchanger for trace phosphate removal from water and waste water. *Water Res.* **2007**, 41, 1603–1613.

63. Stevenson, F. J. *Humus Chemistry: Genesis, Composition, Reactions.* Wiley, New York, 1994.

64. Prakash, P., SenGupta, A. K. Selective coagulant recovery from water treatment plant residuals using Donnan membrane process. *Environ. Sci. Technol.* **2003**, *37*, 4468–4474.

65. DeMarco, M. J., SenGupta, A. K., Greenleaf, J. E. Arsenic removal using a polymeric/ inorganic hybrid sorbent. *Water Res.* **2003**, *37* (1), 164–176.

66. Puttamaraju, P., SenGupta, A. K. Evidence of tunable on-off sorption behaviors of metal oxide nanoparticles: Role of ion exchanger support. *Ind. Eng. Chem. Res.* **2006**, *45*, 7737–7742.

67. Cumbal, L. *Polymer Supported Inorganic Nanoparticles.* PhD Thesis, Civil and Environmental Engineering, Lehigh University, Bethlehem, PA, 2004.

68. Cumbal, L., SenGupta, A. K. Arsenic removal using polymer-supported hydrated iron(III) oxide nano particles: Role of Donnan membrane effect. *Environ. Sci. Technol.* **2005**, *39*, 6508–6515.

69. Cumbal, L., Greenleaf, J., Leun, D., SenGupta, A. K. Polymer supported inorganic nanoparticles: Characterization and environmental applications. *React. Funct. Polym.* **2003**, *54*, 167–180.

70. Helfferich, F. *Ion Exchange.* McGraw-Hill, New York, 1962.

71. Donnan, F. G. Theory of membrane equilibria and membrane potential in the presence of non-dialysing electrolytes. A contribution to physical-chemical physiology (English translation). *J. Membr. Sci.* **1995**, *100*, 45–55.

72. Sarkar, S., SenGupta, A. K., Prakash, P. The Donnan membrane principle: Opportunities for sustainable engineered processes and materials. *Environ. Sci. Technol.* **2010**, *44*, 1161–1166.

73. Brady, P. V., Walther, J. V. Surface chemistry and silicate dissolution at elevated temperatures. *Am. J. Sci.* **1992**, *292*, 639–658.

74. Li, P., Sengupta, A. K. Intraparticle diffusion during selective ion exchange with a macroporous ion exchanger. *React. Funct. Polym.* **2000**, *44*, 273–287.

75. Crank, J. *The Mathematics of Diffusion*, Second Edition. Clarendon Press, Bristol, England.

76. Rawat, J. P., Singh, D. K. The kinetics of Ag+, Zn2+, Cd2+, Hg2+, La3+, and Th4+ exchange in iron(III) antimonate. *J. Inorg. Nucl. Chem.* **1978**, *40*, 897.

77. Srivastava, S. K., Bhattacharjee, G., Tyagi, R., Pant, N., Pal, N. Studies on the removal of some toxic metal ions from aqueous solutions and industrial waste. Part I. Removal of lead and cadmium by hydrous iron and aluminum oxide. *Environ. Technol. Lett.* **1988**, *9*, 1173.

78. Gupta, V. K. Equilibrium uptake, sorption dynamics, process development, and column operations for the removal of copper and nickel from aqueous solution and wastewater using activated slag, a low-cost adsorbent. *Ind. Eng. Chem. Res.* **1998**, *37*, 192–202.

79. Blum, A. E., Lasaga, A. C. The role of surface speciation in the dissolution of albite. *Geochim. Cosmochim. Acta* **1991**, *55*, 2193–2202.

80. Dan, R., Scheidegger, A. M., Manceau, A., Curti, E., Baeyens, B., Bradbury, M. H., Chateigner, D. The uptake on montmorillonite. A powder and polarized extend X-ray adsorption fine structure (EXAFS) study. *J. Colloid Interface Sci.* **2002**, *249* (1), 8–21.

81. Strawn, D. G., Sparks, D. L. The use of XAFS to distinguish between inner and outer sphere lead adsorption complex onto montmorillonite. *J. Colloid. Interface Sci.* **1999**, *216*, 257–269.

82. Dimitrova, S. V. Use of granular slag columns for lead removal. *Water Res.* **2002**, *36*, 4001–4008

83. Kosmulski, M. *Chemical Properties of Material Surfaces*. Marcel Dekker, New York, 2001.

84. Kosmulski, M. pH-dependent surface charging and points of zero charge III. Update. *J. Colloid. Interface Sci.* **2006**, *298*, 730–741.

85. Skoog, D. A., Holler, F. J., Nieman, T. A. *Principles of Instrumental Analysis,* Fifth Edition. Brooks Cole, Crawfordsville, MD, 1998.

86. Martell, A. E., Smith, R. M. *Critical Stability Constants*, Vols. 1–6. Plenum Press, New York, 1977.

Index

Printed in the United States
by Baker & Taylor Publisher Services